D0384900

The Shell Makers
Introducing Mollusks

THE SHELL MAKERS INTRODUCING MOLLUSKS

ALAN SOLEM, Ph.D.

Curator of Invertebrates,
Field Museum of Natural History, Chicago

Drawings by

CAROLE W. CHRISTMAN

A WILEY-INTERSCIENCE PUBLICATION

JOHN WILEY & SONS, New York • London • Sydney • Toronto

Library of Congress Cataloging in Publication Data:

Solem, George Alan, 1931–
The shell makers.

"A Wiley-Interscience publication."
Bibliography: p.
1. Mollusks. 2. Mollusks—Evolution. I. Title.

QL403.S64 594 73-20315

ISBN 0-471-81210-2

Printed in the United States of America

10 9 8 7 6 5 4 3 2 1

To Barbara, Anders, Kirsten, and Fritz

To Barbara, Andrea, Kirsten, and Fritz

Preface

Many popular books have been published on *shells* but very few on *mollusks*. References for shell collectors, many handbooks on the shells of some geographical area, and collections of artistic photographs often fail to include even hints that the shells were made by animals. Only technical publications, including a few monumental tomes and thousands of scattered research articles, reveal the molluscan animals. The enthused amateur shell collector does have a few books that give anecdotal observations concerning marine mollusks, but the world of land mollusks in particular is a subject hidden even to most professional biologists.

Man depends on the living world. Concerns about the environment fill the daily press. Under the leadership of conservationists, the *study* of organisms for enjoyment and pleasure receives more emphasis than *collecting*. This book introduces the second largest group of animals, the 100,000 living species of mollusks, but not from the standpoint of human utilization, nor from the question of rarity and desirability to the collector. What is presented herein is the idea that each kind of mollusk is an integrated functioning unit, a living organism. The basic patterns of these units, how they may have evolved, and what problems and opportunities they faced during their evolution are the main themes. These data are presented from the viewpoint of "ecological anatomy." Major evolutionary advances have required shifts in both ecology and structure. Analysis of the coordination between these factors is a powerful tool in understanding the basic patterns of living diversity.

I present new ideas about what the first mollusk might have looked like and why it eventually had a spiral shell (Chapter III). I speculate on the origin of snails (Chapter VI), why pulmonate snails are so successful on land (Chapter X), and why so many land snails evolved toward slugdom (Chapter XII). Some of these ideas are presented for the first time. Others are abstracted and simplified from technical reports that are currently in press. After the manuscript was submitted, a paper by Charles Stasek,[1] which parallels part of my discussion on molluscan ancestors, was received. Our ideas on the origin of the calcareous shell are similar, but we interpret spiral coiling and torsion quite differently.

This book is also for enjoyment and browsing. Shells have an intrinsic beauty. However, the animals that make the shells are far more beautiful and amazing in their diversified ways of living. Hence the book title and the fact that the color plates are mainly of living mollusks.

In the last few years I have been privileged to pioneer in using the scanning electron miscoscope (SEM) for studying mollusks. The details of form and structure that were previously far below man's limited vision are incredible and beautiful. To share this world I have included ideas and pictures from my recent research to illustrate snail feeding (Chapter IX) and some shell structures (Chapters I and XI).

To highlight ideas and to make them understandable in less than the proverbial 10,000 words require crisp illustrations. The line between sketches and artistic illustrations is sharp. Frequently this is all too obvious in books about animals, where a few good photographs contrast with unimaginative and poor line drawings desperately tossed in because no suitable photograph was available. Much of the spark for this book came through working with Carole W. Christman in planning illustrations. Text and drawings were done simultaneously. Her questions and comments brought my text into far more comprehensible form, while the quality of her first factual drawings inspired me to challenge her imagination and creativity. The decision to include discussions on hypothetical mollusks, the origin of snails, and how slugs arose came because I was confident she could translate my ideas into illustrations. I was not wrong. Unless specified, all line drawings are original diagrams or from specimens in the Field Museum of Natural History. Where illustrations are derived from earlier works, "after" means a close copy and "modified" means a free adaptation.

[1] In *Chemical Zoology,* Vol. VII, *Mollusca* (1972).

Science is cumulative. Most facts presented here were first recorded by others. My research is mainly on land mollusks, so these sections contain many more original observations. Where a section depends heavily on the ideas of one person, an acknowledgment is given, but I have made no attempt to annotate sources. For stylistic reasons, I have omitted perhaps a thousand or more "probably, possibly, generally, usually, maybe," and other useful weasel words. Including the scientists who named new species would serve no purpose. Where generic names are sufficient to make a point, species names are omitted.

Over the years several zoologists have been responsible for channeling and expanding my interests. The late Fritz Haas, K. P. Schmidt, E. R. Dunn, Henry A. Pilsbry, and H. B. Baker at crucial times influenced my career beyond measure. I hope they would be pleased with this result of their help and interest. Thanks also are due to Field Museum of Natural History for years of support in my research and collection building activities, and for generous permission to use numerous photographs. Many ideas presented in this book were arrived at during research supported by the National Science Foundation.

Without the help of my wife Barbara, who knew intuitively when my mind was musing creatively, who did not disturb my mental fancies, and who patiently typed my scribbles, this would not have been finished. Anders, Kirsten, and Baden cooperated equally well.

Unless otherwise specified, all photographs are by the author. The SEM photos, reproduced through the courtesy of Field Museum of Natural History, Chicago, were also made by me, with the invaluable assistance of George Najarian and John Lenke of the Electron Optics Laboratory at the American Dental Association Research Institute during cooperative research work with ADA.

For permission to use photographs and drawings, I am deeply indebted to Field Museum of Natural History, Chicago; Don Byrne; Douglas Faulkner; Elizabeth-Louise Girardi; Louise Kraemer; John Wiley & Sons, Inc.; P. J. Darlington; and *The Quarterly Review of Biology*. Permission to quote from Fretter and Graham's *British Prosobranch Molluscs* was given by The Ray Society, London. My gratitude is also extended to Dorothy Karall for her help in the final rush of figure preparation and to Ferdinand Huysmans for printing the many photographs.

ALAN SOLEM

Barrington, Illinois
October 1973

Contents

The Shell Makers
Introducing Mollusks

I
Understanding Variety

The "shell makers" have successfully occupied a wider range of habitats than almost any other group of organisms. They have solved difficult problems of living in an amazing variety of ways. In the process of expanding their ecological horizons, they have diversified into perhaps 100,000 living species. Surveying this incredible diversity within a single volume requires presenting general principles illustrated by carefully chosen examples.

Here I present an introduction to the biology, ecology, and structure of mollusks. Sources for making identifications, references to more detailed accounts of their biology, and a glossary are presented in appendices. I survey basic molluscan patterns, delineate their place in the living world, discuss how various molluscan groups have solved basic problems in living, and present something of their diversity and evolutionary history.

Since the 1600s there has existed an unfortunate dichotomy between *conchology,* the study of shells, and *malacology,* the study of both shells and their makers. Shell collectors and biologists have traveled on parallel but separate tracks of interest, collectors aware of but not understanding the diversity, and most biologists unaware of the extent of this diversity. This book attempts to bridge the gap. Its contents and organization reflect the growing interest by amateurs in observing mollusks as living animals, and by biologists who are starting to recognize unique molluscan research potentials.

MOLLUSKS AND MAN

Practical interest in mollusks as food for man and as objects incorporated or depicted in art and religion far predates any scientific study. People engaged in molluscan commerce long before the dawn of recorded history. Seashells were traded from tribe to tribe, ending hundreds of miles from the ocean as offerings in funeral mounds. Kitchen middens left by seashore dwellers, heaps of empty land snail shells found in archaeological digs in Iraq, and hill sized mounds of Ohio River mussels eaten by Indians testify as to how long mollusks have been part of man's diet. Art objects from cultures in all parts of the world have incorporated shell motifs into designs. Both the ancient Phoenicians and Amerindian tribes in Western Central America used glands from muricid snails to prepare a brilliant purple dye, known by classicists as Tyrian Purple.

For thousands of years seaside visitors have been enchanted by the shapes and colors of shells washed up by the tides or storms. Many shells are taken home as casual souvenirs, but for some visitors the first casual interest expands into a lifelong hobby. The antiquity of shell collecting as a hobby is suggested by the find of some shells in the ruins of Pompeii, the city buried by an eruption of Mt. Vesuvius in A.D. 79. Several of these shells could only have come from the Red Sea, while others were common Mediterranean seashells. Admired for their beauty, often coveted for their rarity, and nearly infinite in their variations, shells of mollusks have been and remain one of the most popular objects to collect.

In spite of this popularity, however, perhaps most people do not realize that these shells were made by animals. In many parts of the Middle East, land snail shells are thought to be snake pillows, inanimate objects never inhabited by an animal. Sophisticated city dwellers show equal amazement at photographs of living marine snails. Most mollusks are very secretive creatures, often less than 1 inch in size, and frequently nocturnal in their activity. They do not bite like mosquitos, buzz like flies, or scamper through walls like mice. Rarely do mollusks call human attention to themselves.

Mollusks do affect man, and are used in many ways. Gourmets the world over treasure the slippery feel of a raw oyster or the hearty taste of cooked clams. Countless canaries have pecked at cuttlefish bones. Shell jewelry is worn daily. Ground fossil clam shells become lime for chicken feed or are scattered as road gravel. Small spheres of Ohio River mussel shells are inserted into the bodies of pearl oysters off the coast of Japan. A decade or more later some of these pellets are sent back to the United States as the nuclei of cultured pearls. Tiny snails glide on the muddy bottoms of Nile River rice paddies. They carry young worms that later will enter the bodies and sap the strength of Egypt's farmers. Gardeners the world over mutter about the damage done to ornamental plants by snails and slugs.

These are a few of the more obvious and direct ways in which mollusks affect mankind, and are the only reasons most people are aware that mollusks exist. Yet an acre of damp hardwood forest can have 4,500,000 to 11,500,000 snails quietly scraping away at decaying leaves and bits of fungi. An acre of open meadow may have 566,000 snails and 87,000 slugs. Riffle areas of some streams and nutrient laden bubbling springs may be literally carpeted by living clams that quietly pump water and filter food particles. Also present will be thousands of algae covered snails that browse on plants and bits of organic debris. Snails, clams, and chitons in wondrous array compete with each other and many different organisms for space on rocky shores. Offshore storm winds bring millions of mollusks from the West Florida mud flats and strand them on the beach. They subsequently die in the open air, if not first eaten by gulls or picked up by beachcombing shell collectors.

Mollusks on land play a major role in recycling organic matter back into simple chemicals that can then be reused by plants. In both fresh and ocean waters, clams are major factors in maintaining water purity. They filter out and consume microscopic organisms and particles of organic matter that otherwise could accumulate in huge quantities. Clams, in turn, serve as primary food sources for many fish eaten by man. Snails are active browsing herbivores and carnivores; they eat each other and a huge variety of other organisms. A few are swimming members of the plankton, feeding on microorganisms and being fed on by fish. Squids and octopuses are active predators, sometimes competing with man for food fish, but also serving as a yet barely tapped food resource from the oceans.

If mollusks were removed from the food chains carrying on the great cycle that circulates matter and stored energy through the living and nonliving worlds, serious dislocations would result. Despite their mostly inconspicuous ways, mollusks play an important and large role in the economy of nature.

THE LARGE AND THE SMALL

No accurate census of the approximately 100,000 species of living mollusks exists, much less statistics on their sizes, shapes and variations. There are simply too many mollusks and too few people studying the living species, to say nothing of the more than 20,000 fossil species discovered to date.

Nevertheless, it is possible to make a reasonably accurate estimate that more than 80% of the molluscan species are less than 2 inches in maximum shell size, and that perhaps only 5% exceed 3 inches. The giant squid *Architeuthis harveyi* is by far the largest mollusk. Its body length may reach 12 feet. Its two longest tentacles are capable of stretching 48 feet in front for a maximum total length of 60 feet. One of these squids might weigh 500 to 600 pounds. In contrast, the smallest known cephalopod is either the $\frac{1}{2}$ inch long male of *Argonauta argo* or the tiny cuttlefish *Idiosepius*.

A tiny nuculid clam recently found in shallow waters off Yucatan matures at less than $\frac{1}{50}$ inch, while the largest known specimen of this species is exactly $\frac{1}{50}$ inch long. At the other extreme for clams, specimens of *Tridacna gigas*, a common inhabitant of Indo-Pacific coral reefs (see Chapter V), can reach 54 inches in length with a shell weight of 507 pounds. The heaviest known shell of this giant clam weighed 579 pounds, but was only 43 inches long. No one has any idea of how much the animal that made this shell would weigh. Compared with the nuculid, the largest *Tridacna* has about 2,000,000,000 times as large a body volume.

Snails vary from the comparative giant *Syrinx aruanus*, a $27\frac{3}{4}$ inch inhabitant of the Great Barrier Reef off Australia, to the $\frac{1}{37}$ inch *Ammonicera japonica*, which lives between grains of sand on beaches of the Japanese islands. Several chitons are about $\frac{1}{4}$ inch in maximum size, but only the giant *Cryptochiton stelleri* from Western North America reaches 14 inches long. Other groups of mollusks show less extreme size variations.

Age is equally variable. Some snails found in temporary ponds can lay fertile eggs two weeks after hatching, while a European freshwater clam, *Margaritifera margaritifera*, may live more than 116 years.

MOLLUSCAN NAMES

Organizing information about mollusks and arranging collections of specimens are possible only by means of a simple and consistent information retrieval system. This is provided by using the scientific names of species and by classifying species in a "taxonomic hierarchy." The scientific name identifies a particular species, while the taxonomic hierarchy attempts to indicate the approximate degree of relationships between species. This combination of devices serves to group and differentiate. It also has a great advantage in that the terms and their meanings are recognized and understood in every language of the world.

Common names of plants and animals can also either group or differentiate. Oak, maple, hawk, fox, and deer group similar organisms, whereas burr oak, sugar maple, sparrow hawk, red fox, and mule deer identify particular kinds. Such common names involve strong cultural biases. Tree, shrub, flower, grain, weed, annual, fruit, leaf, and nut are all familiar and meaningful plant terms to us. They would seem silly to an Argentinean gaucho. He has over 250 words to describe the color of horses, but all plants are grouped into only four categories: bedding for himself, fodder for his horse, woody plants, and a catchall phrase for everything else. His four word plant vocabulary reflects his needs and values; our vocabulary does the same in indicating uniqueness and relatedness. Few mollusks have widely recognized common names, and in general only scientific names are used here.

Scientific names and the hierarchy of biological classification are concerned with uniqueness and relatedness, but in a more precisely defined way. The structure of a scientific name is simple. The mollusk shown in Plate 1 is called *Nautilus macromphalus* (Sowerby). The generic name (*Nautilus*) is a collective term that shows relatedness, while the specific name (*macromphalus*) indicates a group of distinctive populations (a species) that share a common genetic base and evolutionary history. The name in parentheses (Sowerby) is the person who first described and named this species.

For this information retrieval system to function, it is necessary that the name of each animal be unique. This is accomplished by requiring that (1) the same generic name be used only once in the whole animal kingdom, and (2) the same specific name apply to only one species in a genus. The scientific names of organisms should be stable and unchanging. Unfortunately they do change all too frequently. Some of these changes result from increasing biological knowledge, whereas others reflect necessary corrections of accidental errors.

Since over 250,000 generic and more than 2,000,000 species names have been proposed, some names accidentally have inevitably been proposed more than once. When such error is discovered, the first usage is retained, and the later names must be changed.

The multitude of species alive today can be compared to the outermost twigs on a large tree, and their collective history can be compared to the patterns of tree branching. Unlike a tree, we cannot directly see the branching points in time at which species separated from each other, much less follow the various junctions back to the origin of life. Data from all fields of biological research are used to try and reconstruct the various branching points during the course of evolution and to learn how closely living and extinct species are related to each other.

Scientists express some of these branching points through a system of classification. Each lower level of the "taxonomic hierarchy" represents a progressively later branching point in that line. The seven basic categories can be expanded by adding "super-" or "sub-" prefixes to the intermediate stages. All animals are fitted into this hierarchy. The classifications of the European edible snail, salt water crocodile, domestic dog, and man compare as follows:

	Snail	Crocodile	Dog	Man
Kingdom	Animalia	Animalia	Animalia	Animalia
Phylum	Mollusca	Chordata	Chordata	Chordata
Class	Gastropoda	Reptilia	Mammalia	Mammalia
Order	Sigmurethra	Crocodilia	Carnivora	Primates
Family	Helicidae	Crocodilidae	Canidae	Hominidae
Genus	*Helix*	*Crocodylus*	*Canis*	*Homo*
Species	*pomatia*	*porosus*	*familiaris*	*sapiens*

All four species belong to different orders. Dog and man are more closely related to each other than to a crocodile, since they belong to the same class. Crocodile and man have a greater similarity to each other than do snail and crocodile, since the former pair belong to the same phylum. Positions in such a taxonomic hierarchy define the degree of relationship between

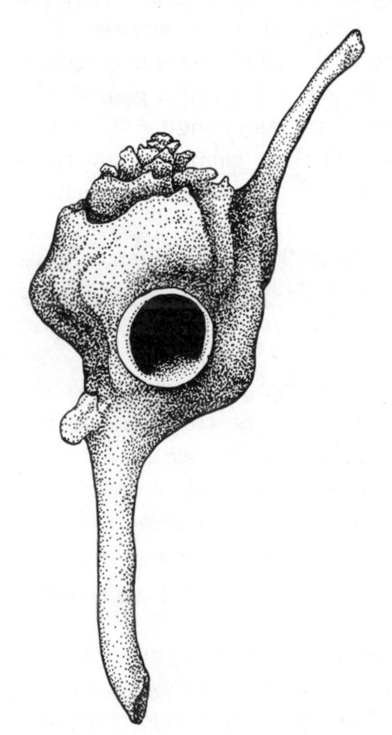

FIGURE 1. *Typhina pavlova,* a deep water Australian muricid snail named for the famous ballerina. After Iredale.

species. The "higher categories" of genus, family, order, class, and phylum represent progressively earlier "branching points" during evolution. These units are based mainly on the observed degree of morphological difference, but they are intended to portray actual genealogical history. This portrayal is far from being perfected. Much of the work by systematists like myself is concerned with trying to understand such relationships between organisms.

The names given to previously unknown organisms are partly at the discretion of the discoverer. The generic name may be predetermined if it is closely related to previously known species, but the specific name can be chosen almost at will, sometimes with intended humor. No malacologist has equaled the opportunity seized by a herpetologist, E. R. Dunn. He named two new salamanders from Mexico *Oedipus complex* and *Oedipus rex.* An Australian conchologist, Tom Iredale, did name a particularly graceful deep water muricid snail *Typhina pavlova* (Fig. 1) after the famous ballerina. Sometimes names are more subtly derived. A rather unremarkable looking New Zealand clam bears the name *Ascitellina urinatoria.* A stream of urine washed out the first specimen of this species from discarded dredged mud in the scuppers of a small boat near the end of a long day at sea.

Each species name will have a story, although usually far less startling or dramatic.

PATTERNS OF KNOWLEDGE

Written information concerning mollusks has been accumulating since the days of Aristotle. His reports on the structure and natural history of Mediterranean marine mollusks include observations that were not verified until early in this century. Thousands of research reports and popular publications have been produced at an ever accelerating rate. Table I presents an estimate of this literature explosion. More than 20 regularly published journals contain nothing but articles about mollusks, and hundreds of other journals will have at least one or two papers on mollusks each year. During the years of the two world wars (see Table I) there was comparatively little research accomplished, but the rapid growth of research publications to the current rate of perhaps 3000 papers each year is impressive. These articles range from short notes on single species to comprehensive tomes of several hundred pages which review a hundred or more species in exhaustive detail, summarize all previously published data, and include many new observations.

The total of about 100,000 titles must be balanced against the 100,000 living and 20,000 fossil species of mollusks. Since covering the anatomy, variation, distribution, biochemistry, ecology, and classification of even one spe-

TABLE I. PAPERS PUBLISHED ON MOLLUSKS

Period	Total	Average per Year
Before 1890 (estimated)	9,000–15,000	
1891–1900	4,907	491
1901–1910	7,129	713
1911–1920	5,432	543
1921–1930	8,210	821
1931–1940	10,093	1009
1941–1950	7,311	731
1951–1960	12,338	1234
1961–1967	16,491	2356
1968–1972 (estimated)	13,500	2700
Estimated total: 94,400–100,400 papers		

cies can require dozens of papers by a variety of specialists, obviously there is much left to learn and record about mollusks. Indeed, many species of mollusks have not yet been named or even collected. Almost every dredge haul from certain ocean deeps, or every visit by a trained malacologist to islands of Melanesia, the tropical jungles of Southeast Asia and Africa, or the Andean forests of South America, will produce previously unknown mollusks. Describing such species is only the very first process in learning about them. Such descriptive work receives less and less attention from malacologists today.

The development of this massive literature has followed a distinctive pattern. Initially there were illustrated books on shells and compilations of descriptions. The floods of specimens brought back by merchant traders and biological collectors on voyages of exploration during the seventeenth through early nineteenth centuries created an excitement and furor almost comparable to that created by the first lunar landing, hard as that may be to realize. Something of the flavor of these times is given in Peter Dance's excellent book *Shell Collecting, an Illustrated History*. By the mid-1800s titles such as *Descriptions of fifty-four new species of land snails* or *Description of thirteen new species of Marginella* were common. Illustrated color plate manuals intended to describe and illustrate all species of mollusks were started independently in Germany, France, Great Britain, and the United States. None of these works achieved their goals, although the number of volumes produced reached 45 in 57 years (Tryon and Pilsbry's *Manual of*

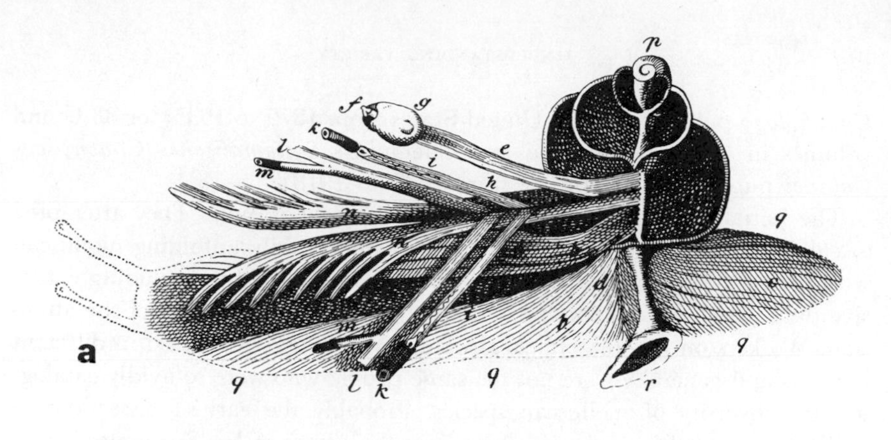

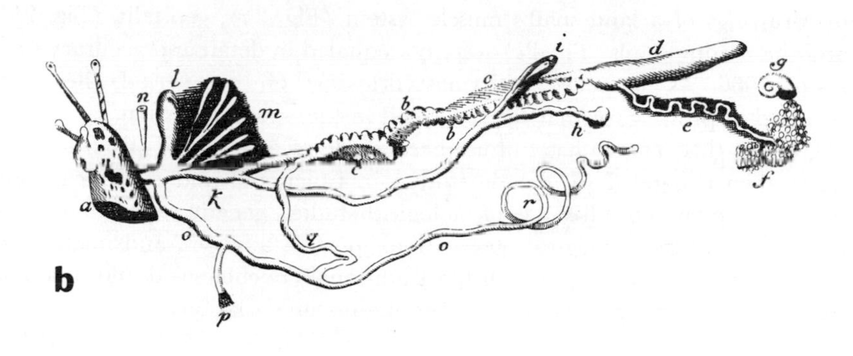

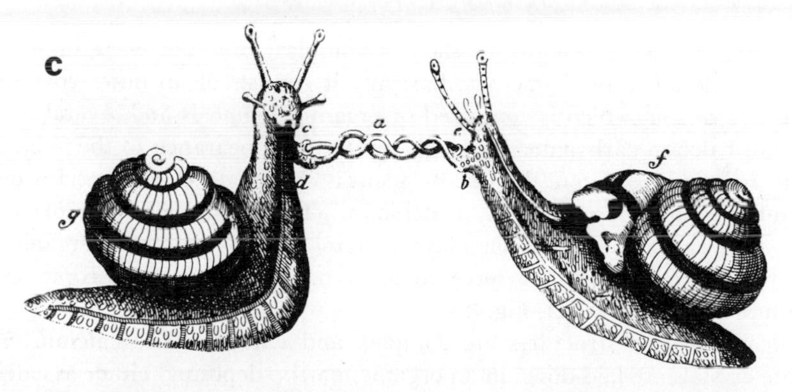

FIGURE 2. Drawings of land snail anatomy made by Jan Swammerdam in the 1660s, but not published in the *Biblia Naturae* until 1737–1738: (*a*) retractor muscle system; (*b*) genitalia dissected out and uncoiled; (*c*) partly separated mating snails. From a 1752 German edition, courtesy Field Museum of Natural History, Chicago.

Conchology published in the United States from 1879 to 1935) or 83 bound volumes in 82 years (Martini and Chemnitz's *Systematisches Conchylien-Cabinet* published in Germany from 1837 to 1918).

The initial impetus of all these works was conchological. They attempted to catalog and display the shell diversity of the rapidly unfolding molluscan world. In time the emphasis in systematic works shifted to including extensive data about the animal, but initially the shell received most of the attention. Workers on the anatomical structure of mollusks published in different journals and generally were not the same people who were so avidly cataloging the diversity of molluscan species. Probably the earliest dissections of mollusks recorded for posterity came from the efforts of Jan Swammerdam in the mid-1600s. Although not published until nearly 60 years after his death, his drawings of a land snail's muscle system (Fig. 2*a*), genitalia (Fig. 2*b*), and the mating snails (Fig. 2*c*) were not equaled in detail and accuracy until about 1900. As happened with most drawings of that period, they were reversed in printing, so that the shells coil in the wrong direction.

Only in this century has full use been made of anatomical structures. Interest then expanded to include studies in functional and ecological morphology. Today the full range of biological studies, genetic experiments, biochemical analyses, ecological observations, protein analyses, and microstructure studies have combined with the traditional conchological and anatomical approaches in seeking to understand molluscan variations.

SHELL STRUCTURE

The structure of the molluscan shell is complex, and the ways in which it can be modified are numerous. Basically it consists of an outer covering, the *periostracum,* which is composed of organic chemicals and several inner layers of calcium carbonate. Although similar in appearance to the "chitin" of the arthropod skeleton, the periostracum is chemically distinct and is usually referred to as "conchin" or "conchiolin." This provides some protection to the underlying calcareous shell layers against erosion or boring organisms. The periostracum also may serve to form much of the microscopic shell sculpture (see Chapter XI, Fig. 8).

Calcareous shell structures are complex and variable. The calcium carbonate crystals are laid down in an organic matrix, deposited either as calcite or aragonite. Calcite is more stable, and thus has a greater probability of being fossilized. It is the main shell substance in oysters, for example. Aragonite comprises the shell of freshwater unionid clams and land snails, while in such archaeogastropod genera as *Haliotis, Nerita,* and *Patella,* layers of calcite and aragonite may alternate.

The exact number and relationships of the calcaerous shell layers vary from group to group. Alternating inner shell layers, such as the "crossed lamellar structures" shown in Fig. 3*a,* are very characteristic just below the periostracum. The basic crystals are laid down parallel to the shell surface, the different angles of deposition presumably serving to add strength to the shell itself. The innermost layer of the shell, that lying next to the soft parts, varies widely. In many marine bivalves and gastropods, it consists of calcite lamellae that lie parallel to the shell surface, giving a dull finish. In special cases such as the apertural barriers of many land snails (Fig. 3*b*), this innermost layer consists of a single crystal layer that is perpendicular to the shell surface.

Most familiar is the sometimes brilliant inner surface of such shells as abalones (*Haliotis*), chambered *Nautilus* (Plate 1), and a variety of marine and freshwater bivalves. The "mother-of-pearl" shell layer in these taxa consists of very fine layers of aragonite, numbering 450 to 5000 layers for each $\frac{1}{25}$ inch thickness, deposited parallel to the shell surface. The crystals are deposited horizontally in plate form, producing an extremely polished surface. Abalones and some freshwater unionid clams, such as *Obovaria,* have an iridescent mother-of-pearl layer, which is caused by physical interactions between the crystals and light waves. Frequently bits of these shells are made into jewelry or embedded in plastic to serve as decorative table tops or trivets.

The relative thickness of the shell layers differs widely both between and within groups of mollusks. Nonmarine mollusks normally have shells with a thick periostracum that provides at least partial protection against acids in leaf litter or during periods when shallow stream or lake waters become acidic because of organic decay. In contrast, many marine mollusks have a relatively thin periostracum, or it can be absent in such groups as the cowries (see Plates 6 and 7). Exceptions to this generalization are many: "Hairy" marine shells and some cone shells (Plate 4) have a thick periostracum, whereas some of the brilliantly colored tropical tree snails (Plates 11 and 12) have a very thin periostracum. In a few carnivorous land snails, such as the New Zealand *Paryphanta* (Plate 12), and in some "semislugs" from tropical forests (see Chapter XII), the periostracal layer is quite thick and the calcareous layer is reduced to a very thin layer or even absent. Both chitons (Plate 2) and the chambered *Nautilus* (Plate 1) have shells in which the shell layering pattern is exceedingly complex.

This brief review of shell structure hints at the great range of variation

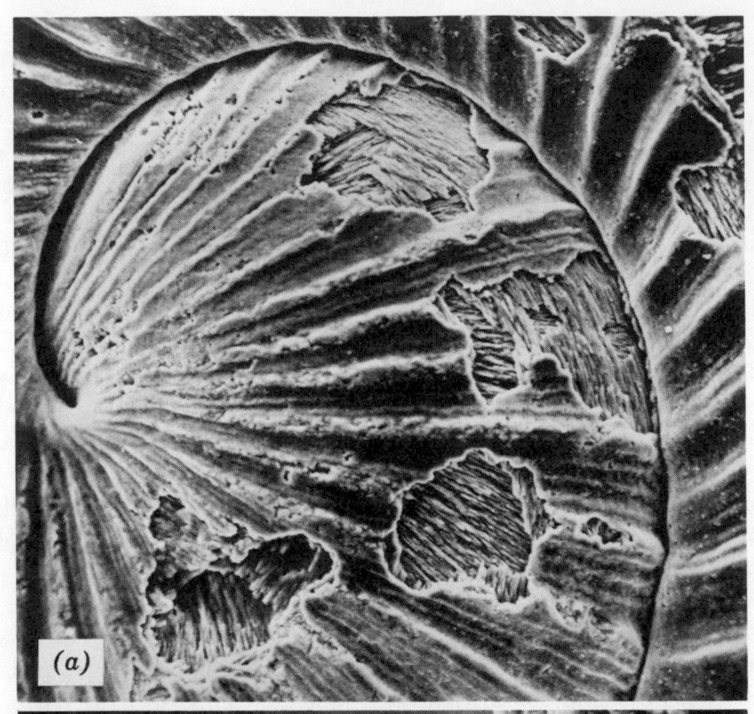
(a)

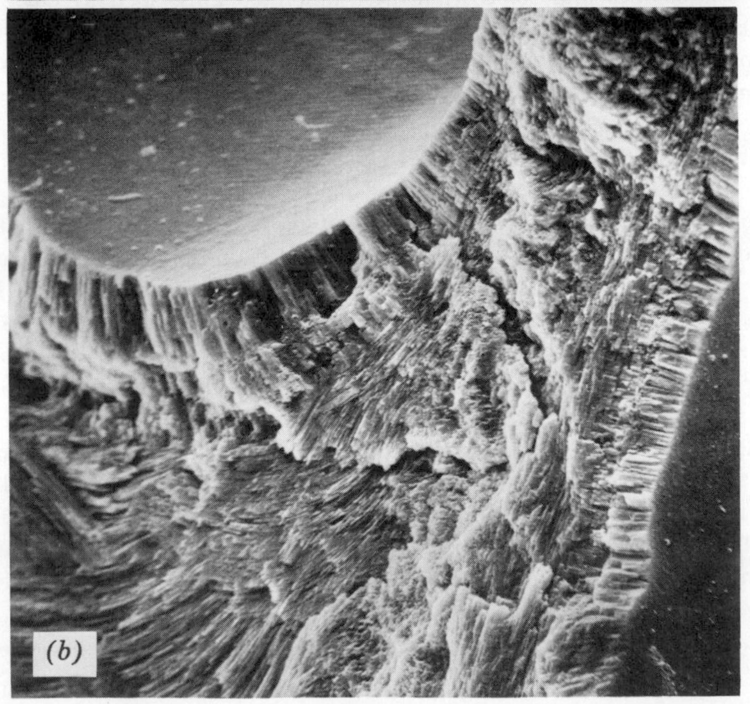
(b)

found in mollusks. It is indeed difficult to make more than a few generalizations about the groups without dwelling on major exceptions. There are a few common features, and the diversity existing today stems from combining these features in a variety of ways. Chapter II, "The Molluscan Patterns," reviews these basic features and discusses how the living classes of mollusks differ in their recombinations. Chapter III, "From Life's Start to Shells," focuses on their ancestors and early history, while Chapters IV through XII explore facets of living diversity. A short discussion of principles relating to current distributions (Chapter XIII) includes recent evidence from the theory of continental drift and ends this introduction to the "shell makers."

FIGURE 3. Shell structure of land snails: (*a*) worn apex of a subfossil land snail from the Gambier Islands, South Pacific showing inner "crossed lamellar layers" of aragonite crystals at 325×; (*b*) fracture section through an apertural barrier (see Chapter XI, Fig. 3*b*) of a Fijian land snail, *Thaumatodon spirrhymatum*, showing the outer perpendicular crystals that form the "mother-of-pearl" shell layer, and the inner lamellar layers that lie roughly parallel to the surface at 1275×. Courtesy Field Museum of Natural History, Chicago.

II

The Molluscan Patterns

Generations of biology students have been puzzled by textbook statements that mollusks are one of the most easily defined and distinctive groups of animals. Obviously a squid, a land snail, a chiton, and a clam are very different not only in appearance but also in their way of life and place of dwelling. It is far easier for students to accept the idea that crabs and insects are related. Both of these arthropods have a hard external skeleton from which sensory organs and mouth parts protrude, and both use long, jointed legs for walking. With a little more study, the basic similarity of vertebrates as superficially dissimilar as fish, frogs, birds, mice, and man becomes apparent. They share both an internal skeleton made up of basically the same bones and the same internal organs. Scales of fish and reptiles, smooth moist frog skin, bird feathers, and mammal fur represent specialized adaptations to different places and ways of living, as do the very different body shapes and relative bone sizes in the various kinds of vertebrates.

Both arthropods and vertebrates have rather rigid, but quite different, body plans. All members of each phylum have a basically common structure. Arthropods have a rigid external skeleton from which sensory and locomotor organs protrude. Vertebrates have a rigid, jointed internal skeleton around which the organs are clustered, and the body is covered by a thin, flexible skin that has the varied surfaces outlined above.

In marked contrast to this, the Phylum Mollusca is characterized by a highly flexible body plan. A very few elements are combined and greatly distorted to enable very different ways of living and methods of locomotion.

In its simplest dimensions a mollusk can be thought of as consisting of a body mass (visceral hump) covered with protective armor (shell) from which extend sensory-feeding (head) and locomotor (foot) areas. They are altered and combined in the various ways that form the confusingly different patterns of the major molluscan groups. Two structures are unique to the phylum Mollusca, the *mantle* and *radula*. The mantle is a flap from the visceral hump. It extends out and downward, thus forming a cavity, naturally called the mantle or pallial cavity, between its outermost edge and the visceral hump. The mantle itself secretes the molluscan shell, increasing it in size by adding to its edge and strengthening it by adding additional inner layers.

The mantle cavity is both the key to molluscan success and the limit to molluscan potential. The external openings of the digestive, excretory, and reproductive systems usually enter into this cavity, which also normally houses the respiratory surfaces, either in the form of "gills" or a "lung" surface. In water dwelling mollusks, the beat of cilia or actual muscular pumping passes currents of water through the mantle cavity, providing respiration (the gills extract oxygen from the water and pass carbon dioxide into the water), eliminating undigestible food matter from the anal opening of the digestive tract, washing out excretory products released from the kidney, and conveying reproductive materials from the animal into the water. Usually there are special organs near the front of the mantle cavity that combine our senses of taste and smell. These organs sample and analyze chemical traces in the water flowing through the mantle cavity. These traces tell the mollusk whether food or a potential enemy lies just ahead. The mantle cavity also often provides the space into which the head and foot of the mollusk can be withdrawn for protection. In squids and octopuses it is specially modified so it can be used as a jet propulsion device.

The other uniquely molluscan organ, the radula, is essentially a "toothed tongue." From a few to a quarter of a million individual teeth are mounted on a flexible membrane. This structure is then partly extruded from the mouth and licked or scraped across the surface on which the mollusk is feeding. Worn teeth are discarded at the anterior end of the radula, while new teeth are formed at the posterior end. The membrane gradually grows forward, moving new teeth into the feeding position and worn teeth to the area where the membrane can be resorbed by the animal. The teeth fall off and may be swallowed with the mollusk's food. Surrounding and operating the radula are a complex mass of muscles. Radular functioning and variation in the tooth structure of snails are discussed in Chapter IX. Clams depart from the molluscan pattern in that the radula is absent. The breathing organs (gills) have become enlarged and greatly modified to filter food particles from water currents pumped through the body.

The molluscan head includes eyes and various chemoreceptors as well as the mouth opening with its radula and associated muscles. Clams have essentially lost their head, since they filter food with the gills in the mantle cavity, sensory organs have shifted to the mantle edge, and there is no radula. Squids and octopuses have the head modified into a circle of flexible arms equipped with powerful suckers.

The molluscan foot undergoes similar alterations. It is basically a broad, flat organ that uses ciliary waves and/or muscular contractions for clinging or gliding. In clams it is either greatly reduced (in those species that are attached to other objects) or modified into a digging organ, while in cephalopods it has become part of the jet propulsion apparatus.

Most other organs of the body are clumped or coiled in the visceral hump, an area elevated above the foot and usually completely and permanently covered by the protective shell. The visceral hump is the center of body metabolism and contains the vast bulk of digestive, reproductive, circulatory, and excretory organs. Technically, the mantle complex can be considered just a part of the visceral hump. Because the mantle and its cavity are unique to the Mollusca and its variations serve in great part to define the different classes, considering it to be a distinct region is desirable.

The shell is probably most accurately viewed as an armored shield for the tender morsels hidden beneath. The shell is formed by the versatile mantle and is extremely variable in both form and ornamentation. Perhaps the most clearly defined trend in shell variation is for it to become reduced in size relative to the soft parts and/or even completely covered by the soft parts. In virtually every major group of mollusks there are some species in which the shell has become reduced to a remnant. In such situations, the animal has evolved other basic defense mechanisms, predation is relatively unlikely, or the energy cost in building a shell is greater than the shell's value.

Different positional and relative size combinations of these basic body features are represented today by the seven classes of living mollusks:

CLASS POLYPLACOPHORA.	Chitons
CLASS APLACOPHORA.	Solenogasters
CLASS MONOPLACOPHORA.	Segmented limpets
CLASS GASTROPODA.	Snails and slugs
CLASS BIVALVIA.	Clams, oysters, mussels
CLASS SCAPHOPODA.	Tooth shells
CLASS CEPHALOPODA.	Squids, octopuses, chambered nautilus

Most solenogasters and all segmented limpets are very rare, deep-sea dwellers that have rarely been seen by biologists. Indeed, specimens of segmented limpets are absent from most of the largest museum collections in the world. The other five classes all occur in shallow water marine habitats, are easily collected, and are much better known.

The diagrammatic blocks in Fig. 1 show the relative positions and estimate the proportionate size of visceral hump, foot, head, mantle, and shell in typical members of the five better known classes. Except for the clam, which is shown in cross section, these are viewed from the right side of the animal. The following paragraphs introduce each of the seven classes briefly, referring back to this figure when appropriate.

Chitons, members of the Class Polyplacophora, are all marine dwellers. They have a very broad creeping foot and a flattened visceral hump. These organs are protected by an articulated shell (Fig. 2) of eight calcareous plates (1) whose edges are formed by and embedded in a flexible part of the mantle called a "girdle" (2). This is very tough, fibrous, and usually adorned with various spicules or scales. On either side of the foot (5) there is a grooved channel in which up to 88 gills (6) lie. This is the mantle cavity, divided thus into two parts by the foot. The head (4) lacks tentacles and eyes, but numerous microsensory organs lie along the shell and girdle surfaces. The digestive tract runs along the length of the body, with the anterior mouth (3) and the anus (7) at the posterior. There are excretory openings on either side near the posterior end of the body. Because of the chiton's elongated form, the mantle cavity, in effect, extends laterally along each side of the foot. Weak currents of water pass from front to back along the gills (respiration), then pick up and carry away excretory products and undigested food matter. The articulated shell and the flexible mantle edging it enable the chiton to cling closely to rough-surfaced rocks. The approximately 650 species of chitons are thus flattened mollusks (Fig. 1) with a split mantle cavity extending laterally along each side of the foot, the head reduced in size, the foot large, sensory equipment diffused along the body, with shell and mantle edge modified for clinging to rocks.

Solenogasters, members of the Class Aplacophora, are wormlike inhabitants (Fig. 3) of marine waters. Most of the perhaps 250 species live at depths of 60 to more than 12,000 feet, although some are found in shallow subtidal waters and a few tiny species form part of the interstitial fauna living between grains of sand. Although most are very small, one species, *Epimenia verrucosa,* may reach 12 inches in length. All species have in common an elongate, rounded shape with the shell reduced to calcareous spicules scattered through the skin. In many species these spicules protrude externally, presumably giving the animal some protection against being bitten. There is no clearly differentiated head, circulatory system and digestive sys-

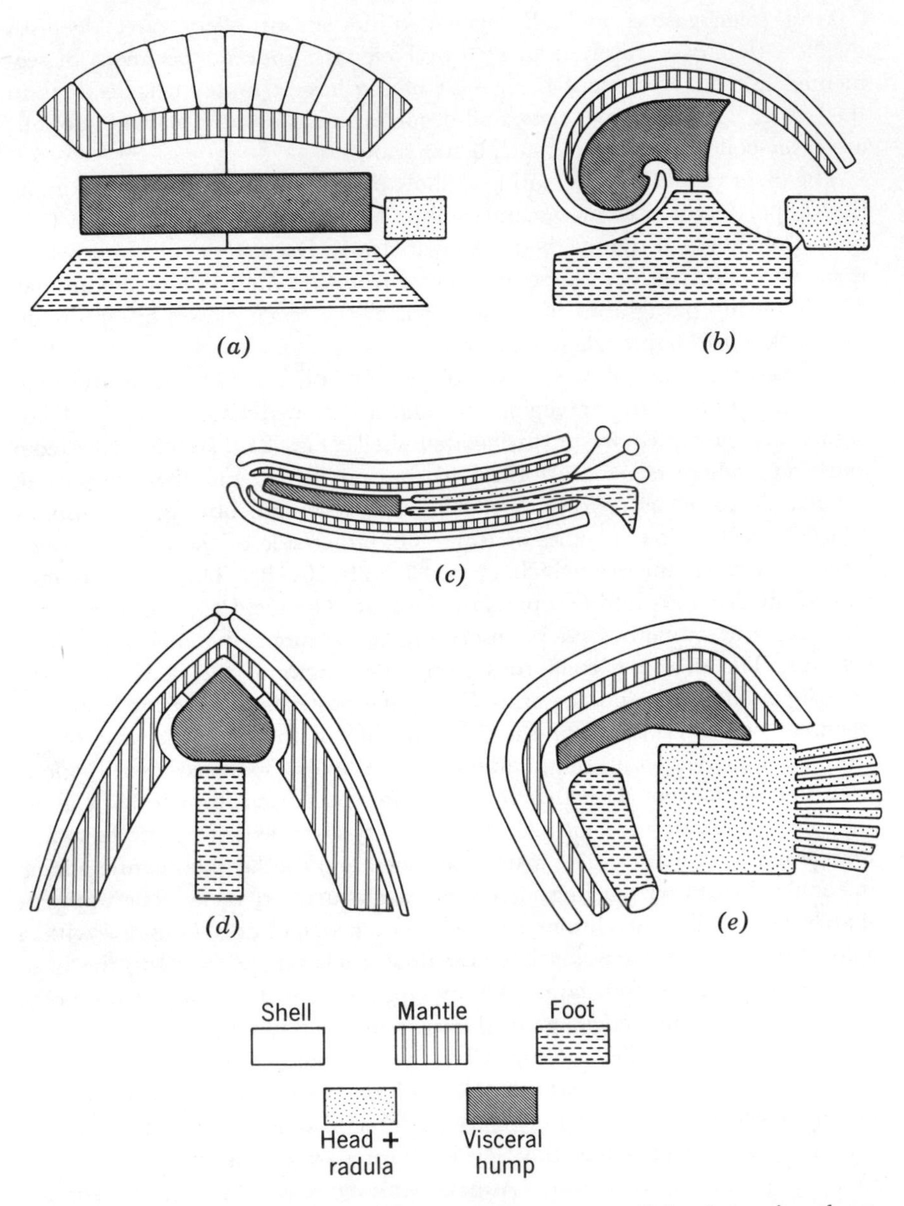

FIGURE 1. Basic parts of the molluscan body as arranged in the major classes: (*a*) chiton; (*b*) snail; (*c*) scaphopod; (*d*) clam; (*e*) cephalopod.

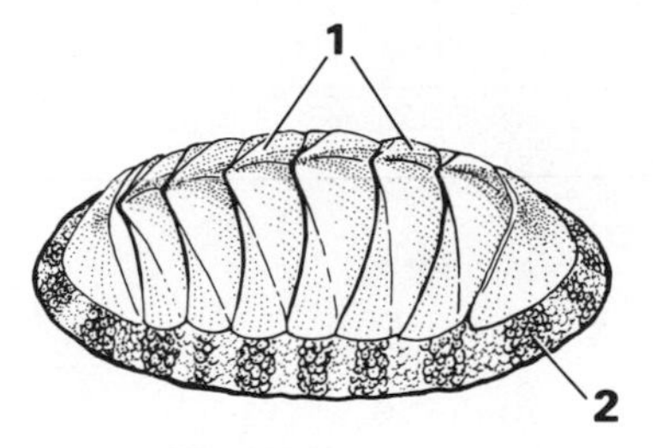

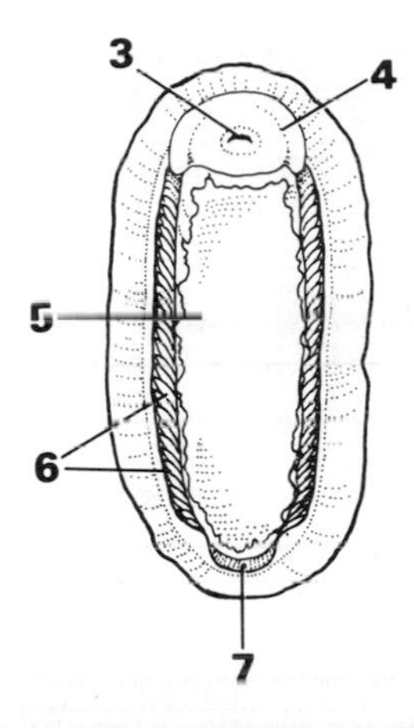

FIGURE 2. Upper and lower sides of a chiton showing shell plates (1), girdle (2), mouth (3), head (4), foot (5), gills (6), and anus (7). Length 2 inches.

tems are simple, and the body contains many bundles of muscles repeated at intervals that extend from top to bottom of the animal. Solenogasters are adapted to two quite different ways of living. The resulting structural adaptations have led some workers to recognize two subclasses or even to divide the Aplacophora into two separate classes. The former approach is more consistent with the level of differences achieved and is followed here. The Subclass Ventroplicida includes forms such as *Dondersia* (Fig. 3) in which there is a narrow but clearly visible foot (2) extending from head (1) to tail, locomotion is by gliding on a mucus trail through muscular contractions, respiration takes place along the lateral grooves of the foot or gill-like structures clustered at the posterior end of the body, and the animal feeds mostly on attached bottom dwelling animals such as coelenterates. The radula is unusual in that the teeth are mounted on epithelial tissue rather than on a separate ribbon as in snails. There may be many teeth in a row, two teeth, or only a single tooth. In some species the radular teeth

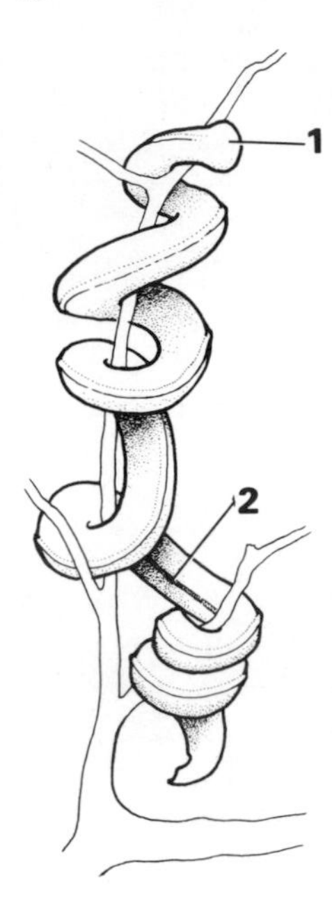

FIGURE 3. The solenogaster *Dondersia* showing head (1) and foot groove (2). Length 1½ inches. After Pruvot.

are greatly reduced in size, but the general presence of a functioning radula, reduced but typical molluscan foot, and the equivalent of mantle grooves alongside the foot are sufficient evidence to relate the Ventroplicida to the other molluscan groups. If only the second subclass, the Caudofoveata, were known, molluscan affinities would be less certain. These species, such as *Chaetoderma,* burrow in muddy bottoms, moving through the muck using much the same type of muscular contractions seen in earthworms. They feed selectively on organic debris and the radula is greatly reduced to absent. Much of the time they sit in a self-constructed burrow, head down, and gills expanded into the water. Probably as a direct adaptation to their specialized burrowing habits, the sides of the body have extended and fused ventrally so that there is no foot surface except for a very short shieldlike area at the anterior end. Respiration takes place through a pair of gills in a posterior cavity. The essential loss of the foot surface, shift in respiratory area, and reduction in radular structures are best viewed as adaptations to the burrowing habitat, rather than as indicating fundamental phylogenetic changes.

For many years, solenogasters were grouped with the chitons, Class Polyplacophora, into a single class called the Amphineura. This was based on the very similar nervous system in the two groups, the development of calcareous spicules in the skin, and details of the sensory, circulatory, and reproductive systems. Chitons and solenogasters probably shared common ancestry in the distant past, but they have diverged into greatly different ways of living and cannot be united into one class. Whether solenogasters are truly primitive mollusks or are secondarily simplified from protochiton ancestors is still being debated by scientists. The typical differentiated molluscan head, foot, shell, and mantle cavity are absent in solenogasters, which suggests to me that they probably are secondarily simplified for their specialized burrowing or browsing patterns of living.

The segmented limpets, or monoplacophorans, are represented by about 10 living species dredged from depths of 6000 to 21,000 feet in several oceans, plus a large and diverse number of species known as fossils from Lower Cambrian to Middle Devonian rocks. For roughly 375,000,000 years they were absent from the fossil record. The announcement in 1957 that a living species of the Class Monoplacophora had been found must be ranked as one of the most startling and significant zoological discoveries of this century. At first glance monoplacophorans are unremarkable. Their caplike shell of $\frac{1}{10}$ to $1\frac{1}{2}$ inches looks exactly like an ordinary limpet. Indeed, the first living specimens, collected in 1952, had been sorted from the dredge haul and grouped with other limpets for study. Monoplacophorans are very remarkable in their anatomy (Fig. 4). There is clear evidence that the body is partly segmented with five or six pairs of gills (4), 10 pairs of lateropedal nerves, five pairs of excretory organs, two pairs of gonads, two ventricles and four auricles, plus eight pairs of muscles attaching the body to the shell. The difference in these numbers is not significant. In both arthropods and annelids the original primitive segmentation has been modified. While an earthworm, for example, retains excretory organs in nearly every segment, it has only five pairs of hearts. Similarly, insects have only three pairs of legs on the thorax, while the many abdominal segments have none. The monoplacophorans prove that mollusks at least experimented with a segmental arrangement. This discovery provides another link connecting mollusks closely to the annelid-arthropod line of invertebrates. The monoplacophorans superficially agree with the chitons in having a broad foot (2) and a mantle cavity extending along each side of the foot, in lacking both eyes and tentacles, and in having a posterior anus (5) and a ladderlike nerve system. Monoplacophorans differ from chitons in showing internal body segmentation, in having a single caplike shell (3), and in virtually all details of structure. The relatively unspecialized head (1) with its lack of tentacles and eyes may simply reflect the problems involved in deep-sea living; eyes and long tentacles would have little value in this habitat. Apparently monoplacophorans feed on very small organisms living in the bottom muck. The laterally extending pallial cavity, broad foot, domed shell

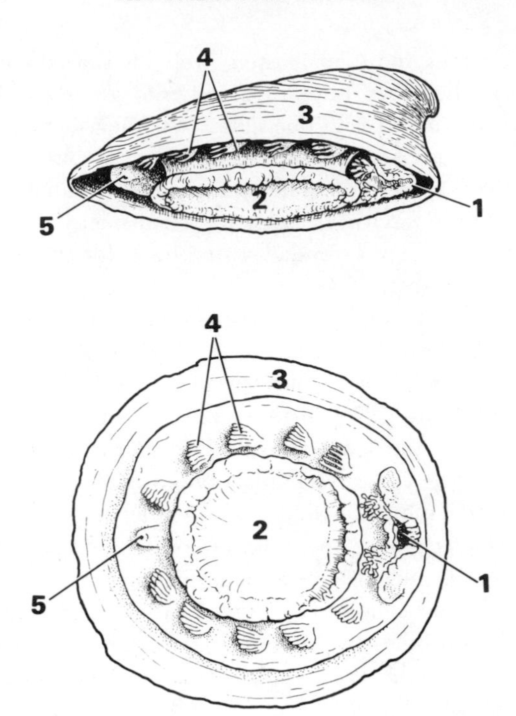

FIGURE 4. The monoplacophoran, *Neopilina,* in side and bottom views showing head (1), foot (2), shell (3), gills (4), and anus (5). Modified from Lemche and Wingstrand.

over the visceral hump, and general form are all what could be called "basic mollusk" in format. The presence of segmentation is the unique feature.

Snails and slugs, members of the Class Gastropoda, have a well-developed head (5) with tentacles, often highly developed eyes (3), a broad foot (1), and usually a high visceral hump (2) covered with what is primitively a coiled shell (Fig. 5). Gastropods are by far the most diverse group of mollusks, with an estimated 24,000 land, 40,000 marine, and 3000 freshwater species. Frequently the shell is reduced in size, altered to a caplike form, or variously elaborated. The most unique aspect of the gastropods is an event called *torsion,* which occurs early in larval life and involves a shift in body orientation. At first the larval gastropod has the mantle cavity at the posterior end of the body, but two 90° rotations of the visceral hump occur at intervals over a short period of time, which brings the mantle cavity to an anterior position just above and behind the head. Not only does this provide a handy empty space into which the head can be withdrawn quickly

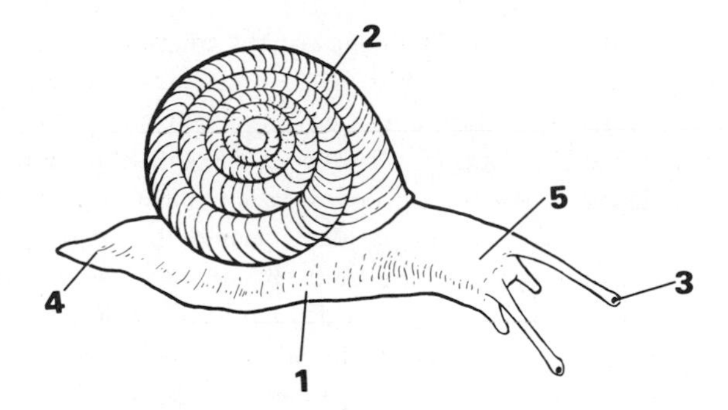

FIGURE 5. An Australian land snail, *Meridolum,* showing foot (1), shell covering visceral hump (2), eyes (3), tail (4), and head (5). Length of foot $2\frac{1}{2}$ inches.

for protection, but in its new position the mantle cavity has the potential for added functions. Chapter VI includes a review of torsion and a hypothesis concerning its origin.

We do not *know* what initiated torsion (see pp. 87 to 93 for one theory of its origin), but in providing both a means of sampling the environment ahead plus an area for quick retreat from danger, the potential evolution of active crawling predators and browsers was enhanced. In time this mantle cavity shift allowed the invasion of land habitats and the great radiation of terrestrial snails. Gastropods represent an elaboration of the head accompanied by specialization of the foot for gliding locomotion on smooth or rough surfaces. Forward rotation of the mantle cavity led to increased efficiency in food location and added protection against disabling injury from a predator. The whole realm of gastropod radiation was opened up by these simple changes. Primitive marine gastropods are mainly browsers on attached animals, while the more advanced species seek out and capture living prey.

An exactly parallel trend developed among terrestrial snails. The more primitive land snails are detritus feeders in leaf litter or feed on algal films, while many of the more advanced species prey on earthworms or other snails. Snails have successfully colonized freshwater habitats as well as the land areas, making them virtually unique among the classes of organisms.

Clams, oysters, mussels, and shipworms are members of the Class Bivalvia. Probably about 15,000 living species are known to scientists. All bivalves are aquatic, with about 1200 freshwater species and the rest marine inhabitants. Bivalves are mollusks in which the mantle cavity has been greatly enlarged in size, and whose gills, in addition to their respiratory function, act as a food sorting organ. In some clams, the gills also serve as a brood chamber for developing young. The edges of the mantle in a clam are partly fused to form a pair of siphons (Fig. 6, 2) that pump and circulate huge quantities of water through the mantle cavity. Elaborate secondary currents caused by cilia pass this stream of water over the gill filaments. Food particles are sorted, passed to the food grooves, and eventually go into the stomach, while nonfood particles are separated, passed to the mantle edge, and eventually ejected to the outside. The shell (3) and mantle (1) have become enlarged to cover the foot (4) and mantle cavity completely. The head region has atrophied, the radula is totally absent, and the body is flattened laterally (Fig. 6). A clam shell consists of two halves or valves, usually equal

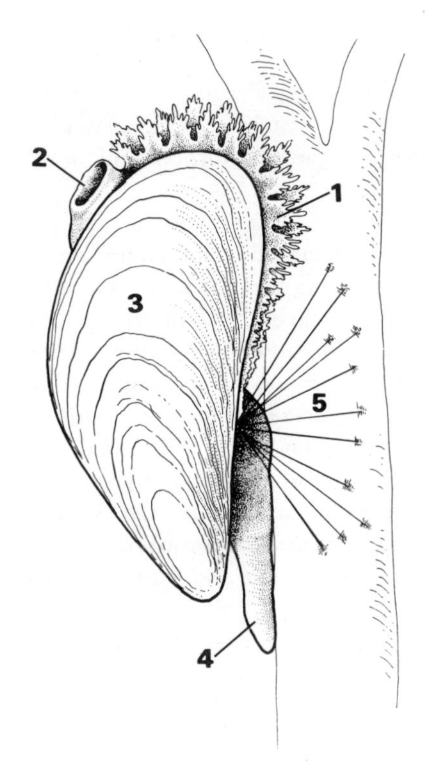

FIGURE 6. *Mytilus edulis,* common edible mussel of Europe, showing mantle (1), one of the siphons (2), shell (3), foot (4), and byssal threads (5) that attach it to the rock. Length 3 inches. After Meyer and Möbius.

in size and shape, that are connected above by a flexible ligament whose resilience keeps them slightly open at the bottom. The valves of the shell can be shut by the contraction of one or two large muscles. The springlike ligament tends to hold the shell open, while the muscles enable the clam to shut its shell, or "clam up." The molluscan foot has been altered into either a digging organ or a secretor of holdfast fibers, since different kinds of clams can be free-living burrowers, remain fastened to one spot during most of their life, or bore into rocks and wood. Bivalves are extremely abundant in shallow ocean waters and several groups have successfully colonized freshwater habitats.

Tooth shells, occasionally called tusk shells, members of the Class Scaphopoda, represent a sideline in molluscan evolution. There are about 350 living species, all of which are found in ocean waters. The shell, which looks like a miniature elephant's tusk, is open at both ends. The narrow end often protrudes above the mud or sand in which the animal lives (Fig. 7). Water currents enter a long mantle cavity that is lined with simple folds

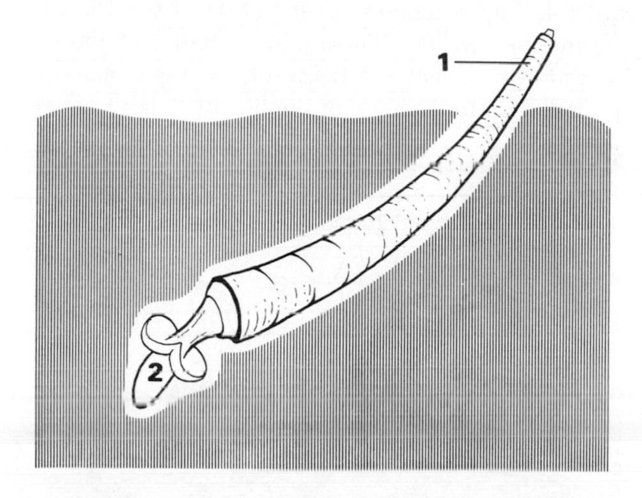

FIGURE 7. A scaphopod mollusk, *Dentalium entalis,* with the shell (1) slightly elevated above the bottom surface and the foot (2) expanded for digging further into the sea bottom. Length $1\frac{1}{4}$ inches. Modified from several sources.

instead of gills. A combination of ciliary action and muscular contractions of the foot circulates water and expels waste materials from the narrow end of the shell. Both head and foot (2) can be extended from the broad end of the shell (1). The foot is modified to serve both in burrowing and feeding; it is used first to raise or lower the animal in the substrate to a point where food is available. Extensions and contractions of the foot pack the sand or mud away from it to create a roughly conically ovate open space or feeding cavity. Long tentacles with expanded clublike disks on their ends, called "captacula," then explore the sides of this cavity, pulling bits of organic debris loose from the wall, or searching out small protozoans called foraminifera and bivalve larvae from among the grains of sand or mud. Collected food items are passed from captacula to captacula back into the mantle cavity and eventually to a series of cilia-covered lips that ring the mouth. A radula with five large teeth in each row then breaks up the food, if necessary, and helps convey it to the stomach. Like gastropods, the scaphopods are torted during development. Elongation of the mantle cavity and tubular shell, specialization of the foot for digging, and development of grasping tentacles on the head characterize scaphopods. In Northwestern North America a common species, *Dentalium indianorum,* was used as money by several Indian tribes, with 25 to 40 large specimens purchasing a slave.

Squids, cuttlefish, and octopuses are members of the Class Cephalopoda. All species are marine dwellers. These active hunters of the midwaters and crevice-ridden shores represent specialization for rapid movement and efficient hunting (Fig. 8). Modification of the head (3) has resulted in tentacles

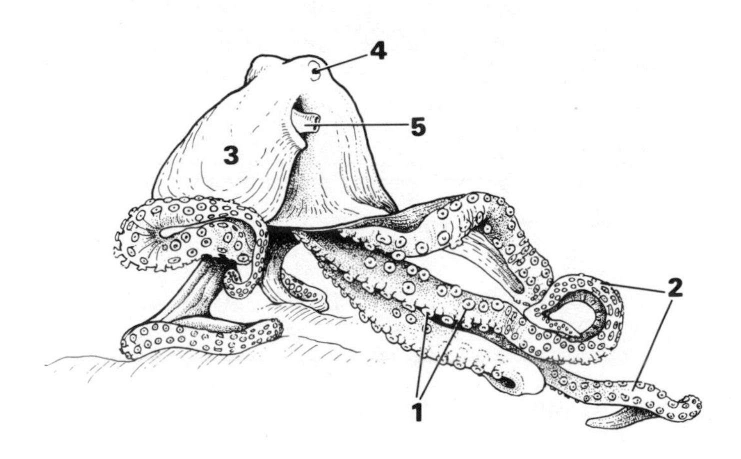

FIGURE 8. An octopus in an aquarium tank drawn so the suckers (1) and tentacles (2) are clearly visible. Head (3), eyes (4), and funnel (5), which opens from the mantle cavity and functions during swimming by jet propulsion, also indicated. Actual length of head about 8 inches.

(2) equipped with suckers (1). Development of a jet-propulsion system (5) from the foot and mantle cavity, reduction of the shell first to an internal fragment (except in the chambered nautilus) and then its subsequent loss in octopods, development of vision (4) that is just as acute as ours, extremely

TABLE II. MAJOR DIFFERENCES BETWEEN CLASSES OF MOLLUSKS

Activity	Polyplacophora (Chitons)	Bivalvia (Clams)	Cephalopoda (Squids, Octopuses)	Scaphopoda (Tusk Shells)	Gastropoda (Snails)
Sensing the outer world	Through their shell	Mantle edge and siphons	Eyes and tentacles	Tentacles	Head and siphon
Food gathering	Browsing on rocks	Filtering	Grab and bite	Grab and scrape	Mostly scrape, some grab, suck, or spear
Food source	Life on rocks	Microscopic life	Larger animals	Microscopic burrowers	Almost everything
Movement	Crawl	Burrow or attached, rarely swimming	Swim by jet or scramble	Burrow	Crawl or swim, some burrowers

high intelligence, and possession of a powerful beak and radula for tearing food are basic characteristics of cephalopods. All of these modifications represent adaptations for free swimming movement and hunting activities. There are perhaps less than 1000 kinds of living cephalopods, although more than 10,000 fossil species have been named. The basic molluscan mantle cavity functions partly as a propulsion device (hyponome), while swift movement and sharp senses substitute for a protective shell. Modification of the head into a food gathering circle of tentacles completed the pattern of cephalopodization.

These then are the basic types of living mollusks. The conservative estimates of specific diversities listed above total about 85,000 species, but the actual total probably exceeds 100,000. By modifying the basic structural units (Fig. 1) a group of slow moving or attached filter feeders (clams), crawling hunters (snails), swimming predators (squids), and rock crawling browsers (chitons) have evolved and radiated successfully. In addition, there still exist three groups that are sidelines in molluscan evolution. These are the tooth shells, which are specialized predators in sediments, the monoplacophorans, a segmented relict from the dawn of molluscan life, and the solenogasters, wormlike predators found at moderate depths.

The classes of mollusks differ not only in structure but also in the ways in which they live. Table II summarizes the patterns of the major classes in respect to how they sense their environment, gather their food, prey on organisms, and move. More detailed accounts of the chitons, cephalopods, clams, and snails are given in Chapters IV through XII.

Studies of the structure, ecology, and patterns of variation shown within the groups of living mollusks have amassed a quantity of data on both their diversity and similarities. There are also common elements in the embryological developments and larval forms of the various classes. By combining all these data and studying the fossil remains of mollusks, many theories and speculations have been developed concerning the origin and early evolution of mollusks. Some aspects of this story are reviewed in the next chapter.

III
From Life's Start to Shells

Mollusks and the myriad other species of organisms living today are the latest products of more than 3 billion years of evolutionary change. Interactions with both the physical environment and the countless other individuals living in the same area go on today. The slow evolution of life proceeds.

Cataloging, charting, and interpreting this diversity have occupied many generations of scientists. This work continues today, but with ever changing emphasis. Starting about 1600, and particularly during the last 200 years, men studied the comparative anatomy of organisms, first at the gross level of whole individuals, then of specific organs, and finally at the level of detailed histology and biochemistry. They recognized varying degrees of similarity and grouped these organisms into the divisions of the taxonomic hierarchy. The many laboriously compiled comparative anatomical studies and systematic summaries of related species groups provided much of the data needed by Charles Darwin to document his brilliant insight concerning the existence and nature of evolutionary change.

Darwin's *Origin of Species,* first published in 1859, convinced the scientific world that evolution was a fact. This conclusion has stimulated more than a century of efforts to trace the lines and branchings of evolutionary change through time. Relationships between living organisms are deduced by studying gross and fine structures, discovering biochemical similarities, finding repetitive patterns of embryological development, making detailed analyses of genetic materials, and surveying patterns of geographical distribution. Direct evidence of evolutionary change comes from the study of fossils, the remains of long dead organisms accidentally incorporated into new rocks instead of being broken down into molecules and recycled into the bodies of other organisms. Sophisticated techniques of recent years have greatly increased our knowledge of the structure and even biochemistry of organisms that became extinct several hundred millions of years ago. Simultaneously, the century old science of ecology, concerned with the structure and functioning of the living world, has focused on the differing roles and patterns of interactions among species.

In the last 20 years a vast synthesis of information from these diverse fields has produced an altered picture of the living world and has greatly extended our knowledge of the earliest history of life on earth.

Where do the mollusks fit into the living world? What organisms have similar structures and seem to be their nearest relatives? What does the fossil record reveal of early life and why are these traces limited? What are the earliest mollusks? Can we suggest how the first mollusks might have evolved? These are the questions with which this chapter is concerned.

BASIC WAYS OF LIVING

Common sense observations led the Greeks to divide the living world into the Plant Kingdom and the Animal Kingdom. Plants grew and reproduced but remained rooted to one spot, whereas animals moved about and reacted to stimuli. The idea of everything being plant OR animal became firmly rooted in Western European philosophy, and it is still taught in most schools today. Scientists discovered much later that plants, through the process of photosynthesis, manufactured organic chemicals. Animals had to get such chemicals by eating plants or feeding upon animals that had eaten plants. This information only seemed to reinforce this classification into two kingdoms.

Yet more than 100 years ago it was known that some microscopic "plants" would manufacture food under some conditions and feed like animals at other times. For many years zoologists preferred to call them plants, while botanists considered them to be animals.

In the last 20 years it has become generally accepted that an organism can obtain organic chemicals in one of three ways: It can be a producer, a consumer, or a reducer. *Producers* take simple chemicals from the environment and manufacture their own organic chemicals. Most producers use sunlight as the energy source, and are what we normally call plants. Producers include the single-celled algae and diatoms found in the waters plus the more familiar trees, mosses, and grasses living on land. *Consumers* obtain organic chemicals by eating producers or other consumers. They have limited capacity to manufacture any organic chemicals. With virtually no exceptions they ingest their food. That is, they take whole organisms, pieces of organisms, or fluids from organisms into their own digestive tract. The higher animals and most protozoans are consumers. The bodies of *reducers* branch or lie in their food supply, which often is the dead or diseased body of a plant or animal. A reducer releases digestive enzymes from its body. They break down the organic chemicals of the "victim" into simple chemicals, part of which are then absorbed into the body of the reducer. Fungi and bacteria are the prime reducing organisms.

The activities of these three types of organisms combine to form a continuous cycle of chemicals from the environment to the bodies of producers who may then be eaten by consumers. The bodies of both consumers and producers eventually are broken down into simple chemicals by reducers, and these chemicals can then be used by other producers.

In addition to the fundamental differences in patterns of nutrition just discussed, many discoveries have been made in the last three decades concerning the body structure of smaller organisms through use of the electron microscope. Whereas studies with optical microscopes had led only to a fundamental differentiation of organisms into "one-celled" and "many-celled" categories, electron microscope examinations of the bacteria and larger "one-celled" organisms such as diatoms and protozoans have revealed that the bacteria have a vastly different and less complex structure than do the others. Many scientists now believe that there are three major levels or grades of structure in the living world. The bacteria and viruses are the smallest and the simplest in structure; the vast number of "one-celled" organisms, perhaps better referred to as "acellular" organisms, constitute a second level of size, ranging generally between 100 microns and 1 millimeter ($\frac{1}{25}$ inch) in maximum dimension; and the "typical plants," animals, and fungi represent a higher level of organization, and attain much larger size. The great size of these latter organisms is possible because their bodies are subdivided into often billions of specialized units (cells) that perform different functions. Through cooperation of these specialized cells, these organisms exist as integrated unitary wholes.

Since about 1860, scientists have been trying to find a better alternative classification of the living world than the traditional "plant" and "animal" kingdoms. By using the three different patterns of nutrition and recognizing

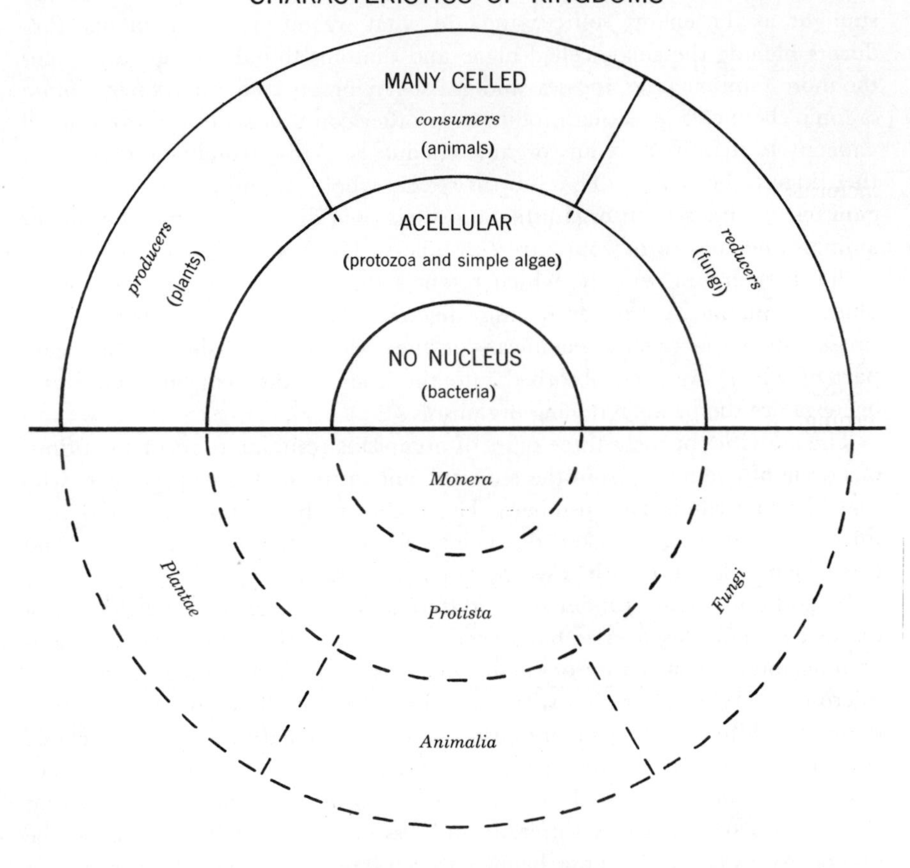

FIGURE 1. Kingdoms of the living world. Lower half of chart indicates the names of the kingdoms; upper half the basic criteria used in recognizing the kingdoms and, in parentheses, the common names of included organisms.

that there are the major structural grades outlined in the preceding paragraph, R. H. Whittaker in 1957 proposed a five kingdom classification (Fig. 1) that has become quite widely accepted:

KINGDOM MONERA. Bacteria, viruses, blue-green algae
KINGDOM PROTISTA. Protozoa, diatoms, golden algae

KINGDOM PLANTAE.	Red, brown, and green algae; liverworts and mosses; ferns; conifers; flowering plants
KINGDOM ANIMALIA.	Sponges, coelenterates, "worms," arthropods, mollusks, echinoderms, vertebrates, man
KINGDOM FUNGI.	Mushrooms, rusts, molds, other fungi, slime molds

This classification gives a much better picture of the structure and functional interrelationships of the living world. The immediately significant fact for this book is that animals form a distinct branch in the many-celled level of organization and that mollusks are one part of this branching. The focus on ways of living provides background data for understanding the role of many mollusks in biological communities.

TYPES OF ANIMALS

The Kingdom Animalia evolved perhaps 700 million to 1 billion years ago from acellular flagellated Protista, probably belonging to the Phylum Protozoa. Major trends within the Kingdom Animalia have been increased body complexity, a tendency toward ever more active hunting for food rather than a more sedentary existence, and increasingly successful colonization of nonmarine habitats. The probable relationships between the major phyla of animals are shown in Fig. 2. This schema departs from standard "phyletic trees" in showing the arthropod and chordate lines as equivalent and parallel, whereas most such diagrams place the chordate groups as representing a higher evolutionary level. At present one species of chordate, man, dominates the earth. But most students of insects would bet (if they could stay around to collect) that insects will be around long after man has disappeared from earth.

These relationships between the phyla have been deduced from comparative anatomy and embryological studies. In particular, the division into arthropod and chordate lines among the higher groups is based on embryological differences and larval structure. Sometimes this dichotomy is obscured by secondary evolutionary changes, but these groups do differ in several basic features. In the arthropod lineage, or Protostomia, the skeleton is mainly external, the early embryological development shows spiral cleavage and very early specialization of the cells, the first swimming larva is of the trochophore type (Fig. 3*a*), and the internal body cavities are formed by splits in blocks of mesodermal tissue. In the chordate lineage, or the Deuterostomia, the skeleton is mostly internal, the early embryological de-

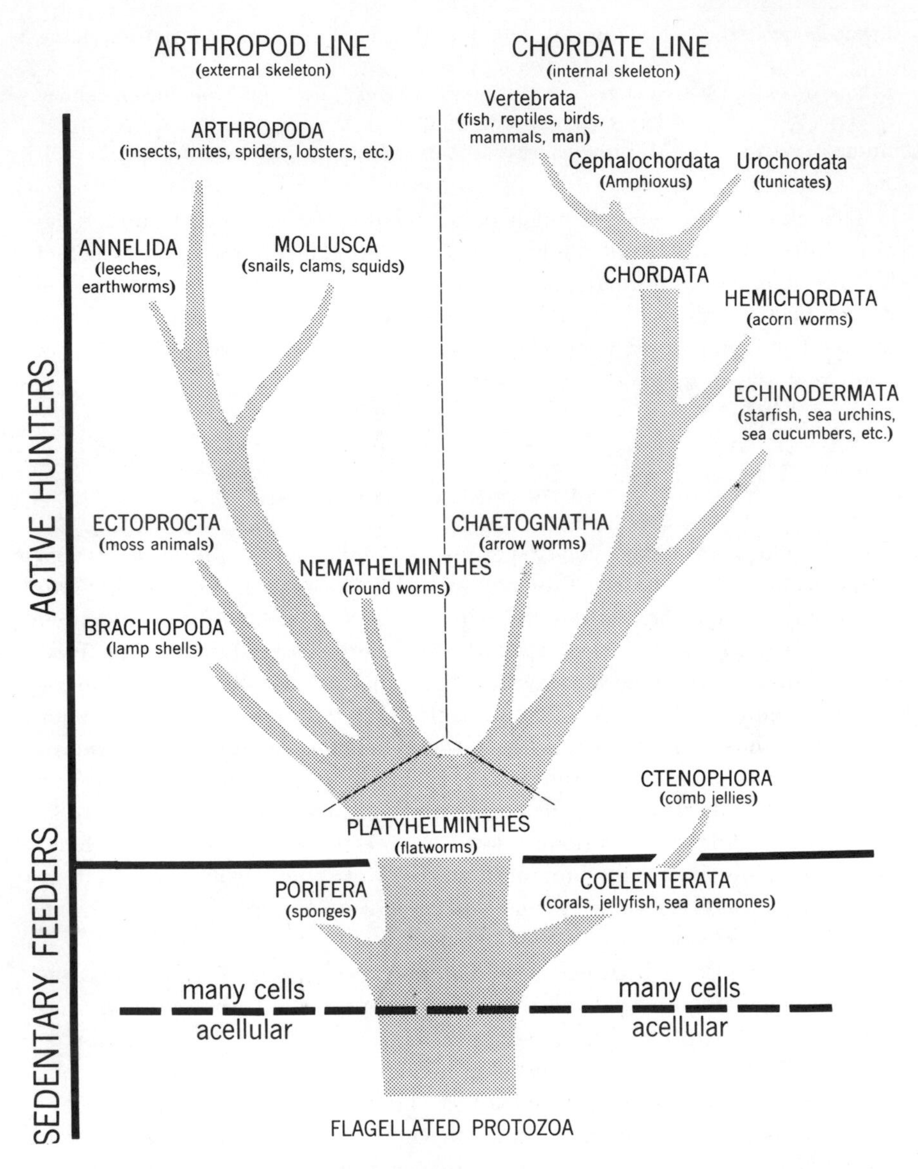

FIGURE 2. Possible interrelationships of living animals, grouped to show major advances in structure and activity. Indicated points of origin and branching are approximate.

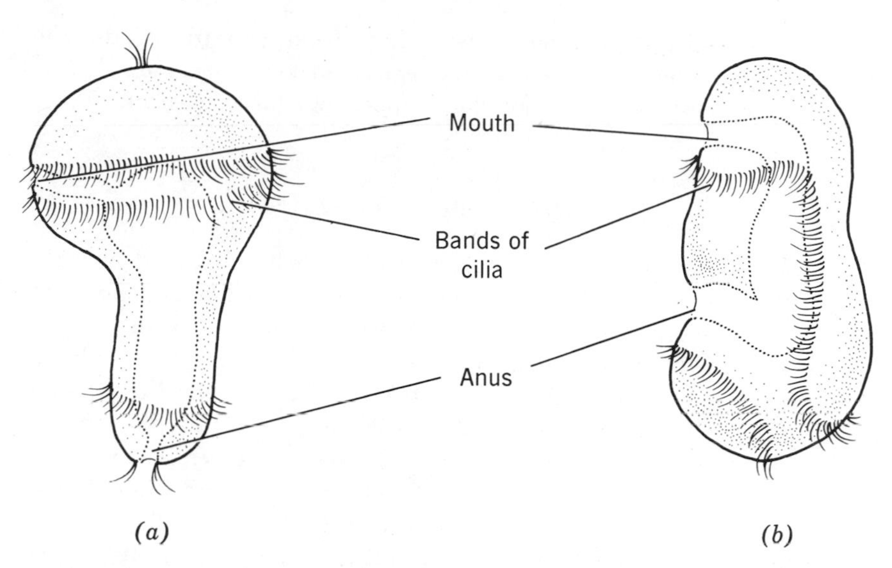

FIGURE 3. Larvae of higher invertebrates: (*a*) annelid type trochophore larva; (b) chordate type dipleurula larva. Both greatly enlarged. Modified from several sources.

velopment shows radial cleavage and specialization of the cells occurs significantly later, the first swimming larva is of the dipleurula type (Fig. 3*b*), and the internal body cavities are formed by pouches from the gut.

The chordate line includes the echinoderms, some of the so-called minor phyla, including arrow worms and acorn worms, the lower chordates (*Amphioxus* and tunicates), and the vertebrates, including man. The highly specialized echinoderms, sea urchins, starfish, sea cucumbers, and crinoids, seem unlikely relatives of man, but their biochemistry, embryology, and formation of an internal skeleton buried in the soft tissue make them the most similar to vertebrates of any phyla. References to introductions on most of these phyla are listed in Appendix B.

At first inspection, the assignment of mollusks to the annelid-arthropod lineage seems improbable. Both of the latter groups have clearly segmented bodies, an external body covering or skeleton from which short locomotor parapodia (annelids) or long jointed appendages (arthropods) protrude, plus complex mouth parts that are used in biting by at least the more primitive species in each phylum. The basic mollusk rasps its food, has a gliding foot, a calcified shell, and shows no evidence of segmentation. The larval mollusk is a modified trochophore. Usually this fact is the feature cited as justifying placement of the mollusks with the annelids and arthropods. The living monoplacophoran, with its partly segmented body (Chapter II, Fig. 4), is another piece of important data supporting this placement.

EVIDENCE FROM FOSSILS

The ancient Greeks and Romans found fossil seashells embedded in rocks on the upper slopes of high mountains. They speculated that these shells might be evidence of past changes in sea levels. During the late Middle Ages these same shells were thought to be "the work of the devil," placed there to confuse man and lead him away from the literal biblical interpretation of creation! In the 1700s and 1800s fossils were accepted as the remains of formerly existing life. Geologists recognized that rocks were deposited in layers, and by comparing and correlating fossils from rock layers in different areas, a picture of change through time emerged. More than 200 years of intensive study has resulted in a fairly detailed picture of life's progression over the past 600 million years.

Unraveling this story was very complicated. Periods of rock building were interrupted by eras in which rocks were being eroded away. Rarely was a sequence continuously revealed in one spot. A few great periods of rock formation and fossil accumulation were separated by equally great periods of erosion. Fossils were used as indexes to tell which rocks were similar in age, and to divide the last 600 million years into the series of eras and periods outlined in Table III. Radioactive and isotope dating techniques have permitted the assignment of actual approximate ages to these divisions.

At the bottom of the table are the Proterozoic and Archaeozoic eras, which cover the far longer time span from earth's origin to the 600 million year mark. They have not yet been divided into generally accepted smaller time units.

Until the last decade, only the sketchiest details were known concerning life in the Archaeozoic and Proterozoic eras. Impressions of a few jellyfish and segmented worms had been discovered in late Proterozoic rocks, but the record is so sparse that most books simply started discussing early life with the Cambrian fossil record. Newer techniques of study and concentrated searches for microfossils have vastly increased our knowledge of pre-Cambrian life. While a plentiful fossil record starts in the Cambrian, traces of life have been found in rocks that date back 3.2 to 3.5 billion years.

There are several reasons for the scanty early fossil record. First, fossils can only be preserved initially in *sedimentary* rocks, that is, those formed from bits of eroded rocks, plants, animals, sand, and clays that accumulate

TABLE III. DIVISIONS OF GEOLOGIC TIME

Era	Period	Millions of Years Before the Present
Cenozoic	Pleistocene	0–1
	Pliocene Miocene Oligocene Eocene Paleocene	1–63
Mesozoic	Cretaceous	63–135
	Jurassic	135–181
	Triassic	181–230
Paleozoic	Permian	230–280
	Carboniferous	280–345
	Devonian	345–405
	Silurian	405–425
	Ordovician	425–500
	Cambrian	500–600
Proterozoic Archaeozoic		600–4500

through the action of wind, water, or glacier movements, and become compacted under pressure into such rocks as sandstones, shales, or limestone. In contrast, *igneous* rocks are formed from molten material, presumably spewed forth from deep within the earth, and thus have no fossils. *Metamorphic* rocks are either igneous or sedimentary rocks that have been changed by heat and pressure when buried under many layers of newer sediments.

All three kinds of rock erode when exposed to the action of wind and water, so fossils can disappear by erosion. In addition, fossils can become warped or be destroyed by the heat and movements in the process of changing a sedimentary rock into a metamorphic rock. The longer the rock has been in existence, the greater is the statistical chance of something having happened to it. So the Paleozoic and Proterozoic eras would be expected to have far less fossil materials available today than rocks from Mesozoic or Cenozoic times simply because of accidents during the long periods since these older rocks were formed.

Second, before the Cambrian, larger organisms apparently lacked skeletons and were thus preserved only through very rare accidents. Most soft bodied organisms have their bodies pulverized by reducers and recycled very soon after death. They must be covered essentially immediately after death, or die because of being covered by sediment. In contrast, bones and shells decay far less rapidly and thus are much more likely to be covered by sediment and incorporated into rocks.

Finally, the vast majority of the period from 600 million to 3.5 billion years ago involved evolution and proliferation of submicroscopic and microscopic life. The time scale in Fig. 4 lists current knowledge of the periods in which major steps in the evolution of life probably occurred. The jump from the Kingdom Protista to many-celled organisms occurred probably less than 1 billion years ago, more than 2.5 billion years after the presumed origin of life. Most of the animal phyla first appeared in the fossil record between the start of the Cambrian and the Middle Ordovician, a span of only 130 million years. The eruption of land life and the rise of mammals, with which most "histories of life" are concerned, came in the last moments of the geologic time scale.

What precipitated the development of skeletons? Possibly it was triggered indirectly by a rise in the free oxygen level (oxygen is a "waste product" of photosynthesis). When enough oxygen was present for efficient respiration by many-celled animals, an explosive radiation of larger sized animals became possible. Certainly a skeleton helps support a large body, and also provides possible protection against being eaten, stores minerals that can be used by an organism, provides a firm point of attachment for muscles that are used in locomotion or for feeding, and also gives a fixed form and shape to an organism.

Skeletons of microscopic animals originated slightly earlier. Many protists, such as foraminifera, radiolarians, diatoms, and dinoflagellates, have microscopic skeletons of almost unrivaled beauty and complexity. These accumulate on the seabeds in such vast numbers that in earlier times they formed the rocks later elevated as the famed white cliffs of Dover in England. Foraminifera skeletons also compacted into the quarried limestone blocks used to build the great pyramids of Egypt.

Skeletons can be made of many substances, silica in the case of radiolarians and many diatoms, the complex organic chemicals classed as "chitins" in the arthropods and segmented worms, calcium phosphate for most vertebrate bones, and varieties of calcium carbonate in many other animals, including mollusks, corals, and foraminifera.

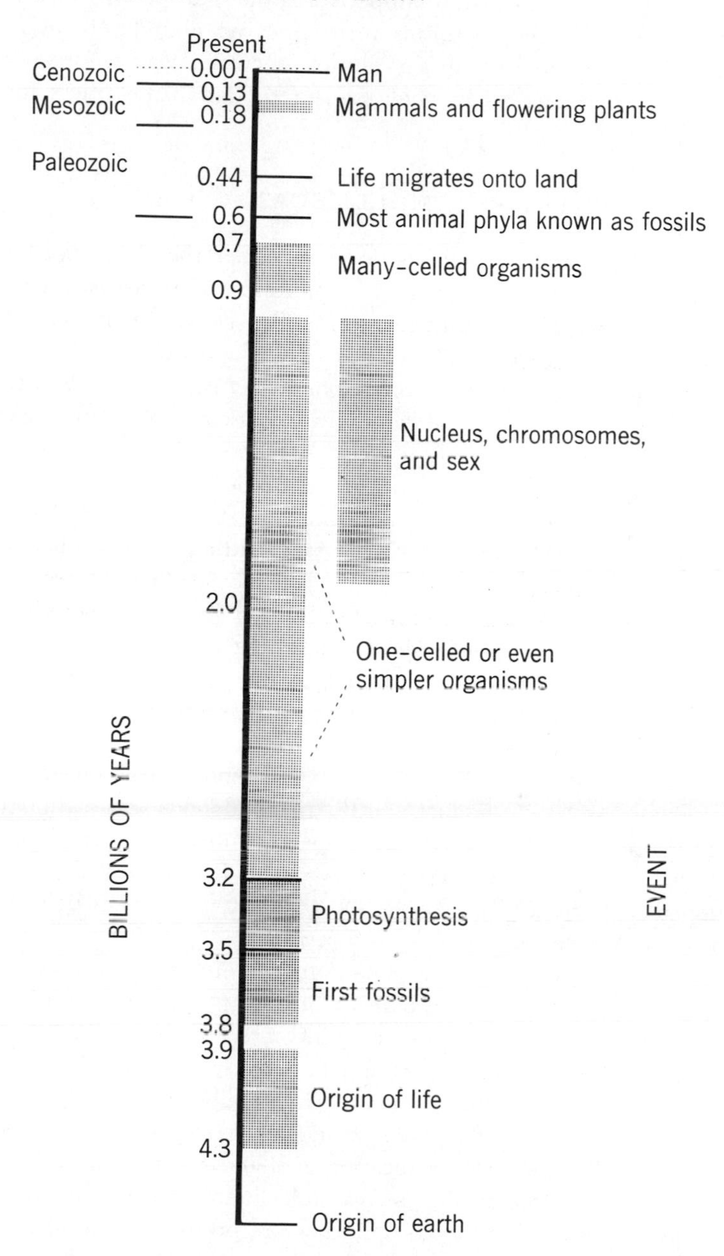

FIGURE 4. Chronological chart of the major events in the evolution of life on earth.

People still think of the Cambrian period as the "start of the fossil record," since skeletons of larger organisms first appeared abundantly in rocks of this age. However, most of life's evolution occurred way before this date, but these earlier organisms left only scattered, tantalizing traces for us to see.

THE FIRST MOLLUSCAN FOSSILS

If fossils of the very first mollusks do exist, either they have not been collected or they are not yet recognized as mollusks. The earliest known fossil mollusks are already recognizable as snails, clams, monoplacophorans, scaphopods, or cephalopods. There are a few fragmentary Lower Cambrian fossils that might be clams or, more probably, crustaceans with a two-part shell. The early fossils that are definitely molluscan can all be assigned to the major living classes.

Monoplacophorans and gastropods are the oldest known molluscan fossils. Both have been found in Lower Cambrian rocks. The earliest monoplacophoran, *Cambridium** from North Asia, had a limpetlike shell showing internal traces of muscle scars on the shell and other lines suggesting organ segmentation. Shells of the two earliest presumed gastropods, both belonging to the now extinct Order Bellerophontina, either have a very simple caplike shell (*Helcionella*) which is very similar to the stage shown in Fig. 5*d*, or a modified limpetlike shell with two whorls (*Coreospira*) which looks only slightly different from Fig. 5*e*.

The first bivalves whose "clammy" nature is undisputed come from lower Middle Cambrian rocks, the first nautiloid cephalopods and scattered valves of large chitons have been found in Upper Cambrian deposits, the earliest records of scaphopods are in the Ordovician of Russia, and the ammonoid cephalopods date back to the Lower Devonian. Only the soft bodied solenogasters, Class Aplacophora, lack a fossil record.

There is no guarantee that the time sequence of these fossils suggests their relative time of evolution. The date of origin of a group and the time of its appearance in the fossil record certainly are not the same. Nor is there any reason to expect that there would be exactly the same length of time between origin and first fossilization. It is equally true, however, that we have no reason to assume that there would be major differences in the time gaps between origin and first fossilization of shallow water marine organisms. Until more definite information is available, it seems reasonable to assume that the gastropods and monoplacophorans are probably more like the "first

* E. L. Yochelson in 1969 referred this to the Class Stenothecoida, but I am not convinced that *Cambridium* had a bivalved shell.

mollusk" than the other classes. Both clams and cephalopods are very highly specialized and thus are assumed to represent subsequent modifications of the "basic mollusk" plan. Scaphopods and chitons are specialized in form and habitat, in addition to appearing later in the fossil record.

Fossils do tell us something of how early the varied types of mollusks appeared. They suggest that the gastropod and monoplacophoran types came earlier than the others, but they give us little information concerning structures of the first mollusks. Data from living mollusks, patterns of evolution in other organisms, and information on ecological relationships between predators and their prey organisms suggest several features of this "protomollusk" and allow speculation on its evolution.

ORIGIN OF MOLLUSKS

Both primitive annelids and arthropods differ from mollusks in having biting mouth parts, locomotion based on appendages, retention of marked body segmentation, and an external skeleton composed mainly of organic matter. The molluscan characteristics involve rasping mouth parts, movement with a broad ciliated foot, little or no body segmentation, and an external skeleton composed primarily of calcium with a thin organic layer on top. The similarities in embryology and early larval development suggest common ancestry, but the major differences suggest that the common ancestor resembled none of them.

It is an axiomatic part of evolutionary theory that major advances and adaptive radiations by organisms involve crossing a threshold into a new "adaptive zone." This adaptive zone can involve such diverse features as exploiting a new source of food, developing a changed pattern of locomotion, or occupying a new habitat. Most attempts at explaining the origin of mollusks have focused on details of structure. Usually this results in constructing an "archetype," a hypothetical ancestor with the common molluscan features of broad foot, paired gills, and excretory organs, a vague head, and simple caplike shell. Discussions and figures in Morton's *Molluscs* and Russell-Hunter's *A Biology of the Lower Invertebrates* are excellent examples of this approach. Each molluscan class is then derived from this ancestor by altering various elements into the equivalent of the block arrangements shown in Fig. 1 of Chapter II.

A slightly different approach is taken here. I focus instead on the basic adaptive thresholds involved in the "molluscan way" as opposed to the annelid-arthropod pattern. Obviously feeding is one of them. Primitive snails, chitons, and monoplacophorans browse on sedentary organisms, using the

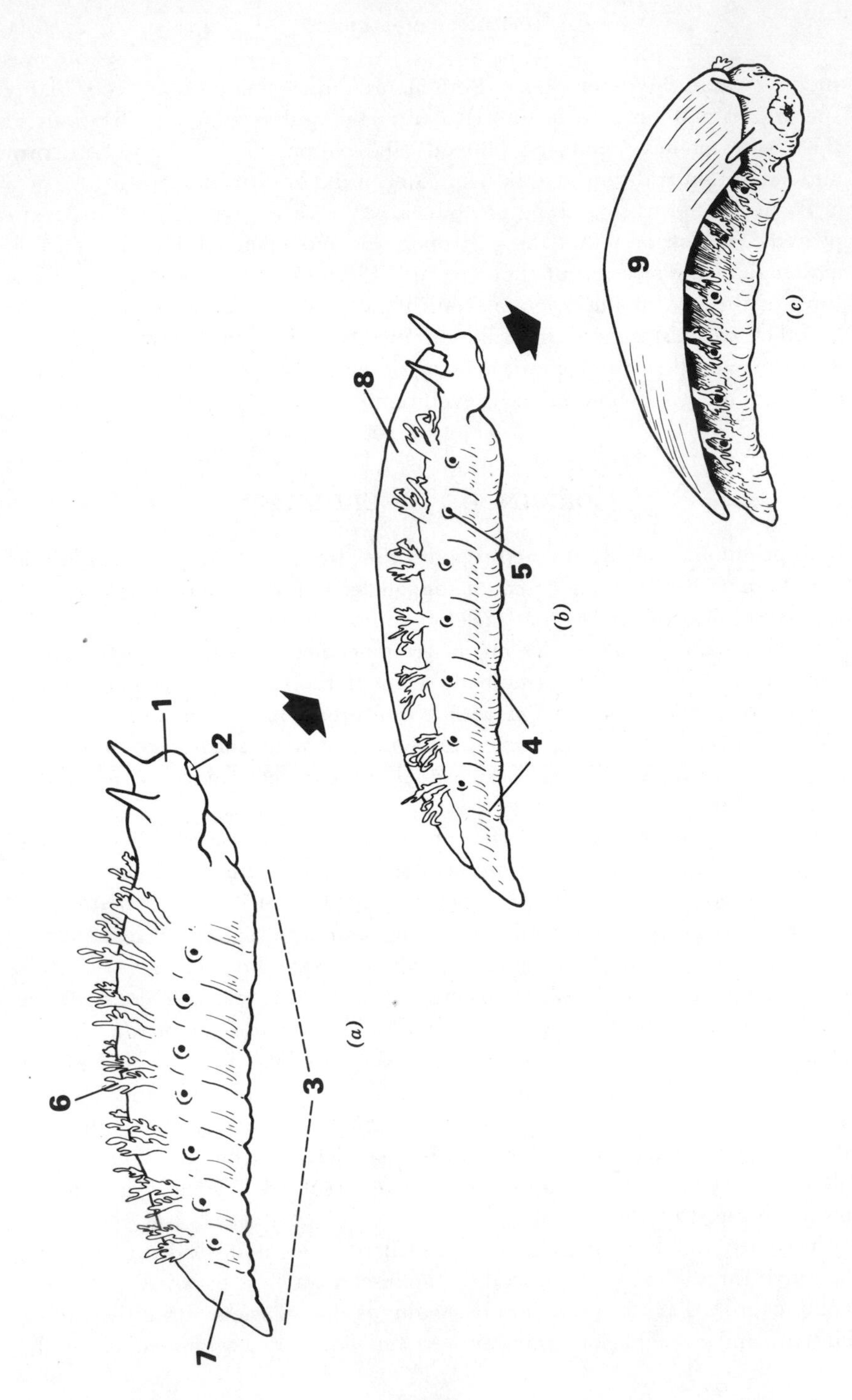
1
2
3
4
5
6
7
8
9
(a)
(b)
(c)

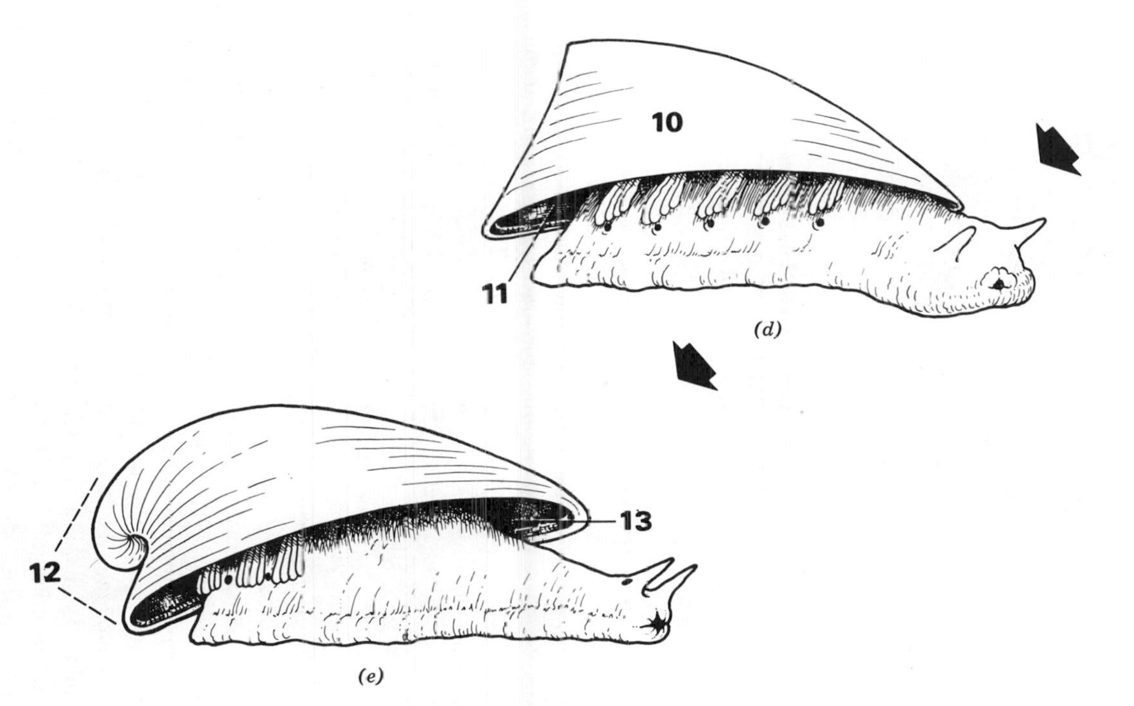

FIGURE 5. Hypothetical early evolution of mollusks from partly segmented wormlike organisms. The successive stages, from top to bottom, are as follows: (*a*) soft bodied ancestor with body large enough to require external gills; (*b*) second stage has a dorsal shield formed of conchin and/or calcium granules; (*c*) development of the shield into a shell plate; (*d*) evolution of a movable shell that can be rocked forward over the head, and the development of a corresponding posterior cavity; and, finally, (*e*) spiral coiling added to the shell with both anterior and posterior cavities. Reduction in the number of gills and external excretory pores probably continued. Labeled structures are head (1), mouth (2), foot (3), segments (4), excretory pore (5), gills (6), tail (7), dorsal shield (8), shell plate (9), shell (10), posterior cavity (11), spiral coiling (12), and anterior cavity (13).

many-toothed radula to scrape small whole organisms or bits or organisms off surfaces. The biting mouth parts of primitive annelids and arthropods suggest the pursuit and capture of larger or more active prey. In order to pursue an active prey, use of appendages for movement with the body slightly to strongly elevated above the surface is highly efficient. In contrast, if the prey is sedentary or very slow moving, the retention of a broad gliding foot as in mollusks and such primitive predators as flatworms (Phylum Platyhelminthes) causes no difficulties. Correlated with the same phenomenon is self-defense against predators. A more agile organism with biting mouth parts can flee and/or try to defend itself by biting. A more slowly moving organism without biting mouth parts must rely on camouflage, chemical defense, or some kind of physical shield, such as the molluscan shell. It thus seems quite possible that the basic split between the annelid-arthropod stock and the molluscan stock started with a divergence in feeding pattern, the mollusks specializing at first on small, stationary food items, and the annelid-arthropod prototype on a more active prey. Subsequently the mollusks developed a variety of active hunters, such as the higher gastropods and the cephalopods. Even the scaphopods and aplacophorans can be considered more active hunters than the primitive mollusks. The clams, in great contrast, became the most successful group of filter feeders, straining nutrients out of the water. But this occurred much later than the initial evolution.

A rather early adaptive change which led to more efficient predators was "completion of the digestive tract." Living flatworms and all coelenterates (corals, jellyfish, sea anemones) must take food into their mouth, digest it inside a cavity that radiates or extends through most of their body, then regurgitate indigestible matter out through their mouth. This system works, but would preclude the evolution of a large, slim, long, active hunter. If, instead, food is taken in at one end of the body and indigestible matter eliminated at the other end, the body can be longer and streamlined for more rapid movement. The roundworms, Phylum Nemathelminthes, have this advance, but generally are very small. A very effective way to elongate structure is to duplicate sections of the body again and again. This provides such services as respiration and excretion to each area by reproducing small organs serially. The other alternative, used by our distant ancestors, is to make the respiratory and excretory organs larger, then use elaborate and lengthy duct work to service more distant areas of the body.

If we start with a flatwormish type of gliding predator, complete its digestive tract, and partly elongate the body through segmentation, then a suitable ancestor for both mollusks and the annelid-arthropod lineage exists. This organism would be bilaterally symmetrical, with a digestive tract opening at the posterior end of the body. As a predator, it would have a moderate concentration of sensory equipment at its anterior (head). Presumably it would move by a combination of ciliary and muscular action, most probably on a broad foot.

Respiration presents problems when confronted with any size increase in other than the narrowest or thinnest of bodies. Small organisms carry on respiration through the whole surface of their body. But the area of a circle increases as the *square* of its radius, while the circumference increases by only *twice* the radius. This means that increase in body surface area is far less than the increased body volume that needs to receive oxygen and expel carbon dioxide. The development of accessory or additional respiratory surfaces protruding from the body, that is, gills, is a logical solution to the problem.

By putting together the features outlined above, a creature very similar to what might have been a common ancestor to both the mollusks and the annelid-arthropod groups is developed. This hypothetical "protoprotostome" is shown in Fig. 5*a*. By switching to biting mouth parts, quicker locomotion through use of appendages to elevate the body, and keeping a relatively thin organic skin or cuticle, this could evolve to the "protoannelid." By specializing in scraping at sedentary animals or algae, developing a dorsal shield of calcium, and retaining a gliding foot, this gradually shifts into a "protomollusk" as diagrammed in the other parts of Fig. 5. The possible causes of these changes are discussed below.

Gills (6) and paired excretory openings (5) often are the external indications of segments (4). The differentiation of head (1) and tail (7) departs from a strict segmentation pattern (Fig. 5*a*). As the animal crawled about seeking prey, its color and upward extending gill filaments probably served as partial camouflage and gave some protection against predation, much as in some living nudibranchs. Quite probably the animal was covered with a thin cuticle of "chitin." Any strengthening of this by deposition of calcium crystals into the organic matrix of the skin would provide some protection against a predator (Fig. 5*b*). If the first bite or grab fails, there is always the chance that the intended victim can escape. Hence development of a dorsal plate (8) would give an advantage for survival. Initially this could have been partly flexible, but with increasing calcification, it would become rigid. The calcium would continue to be covered by an organic layer. This is needed to prevent erosion of the calcium by microorganisms. Growth would have been by additions to the inner surface of this shield. Partial extension of the plate in the form of a flap to cover the head would have been favored, provided this plate was free from attachment to the head in order to permit lateral head movements (Fig. 5*c*).

During early stages the delicate gills would have been exposed to predation. When pointing upward (Fig. 5*a*) they could serve as camouflage and be exposed to clearer water. But development of a hard dorsal shield would tend to shift them more to each side of the body (Fig. 5*b, c*), reduce their camouflage value, and expose them to cloudy water stirred up by the animal's movement. A flap of body tissue on either side could extend the shield edge laterally. In time this would form the mantle. As the shield extended further laterally on each side (Fig. 5*c*), the gills would eventually have to point horizontally or even slightly downward. The increased protection resulting from continued lateral expansion of this protective shield (9) would have high selective value. With the shield extended laterally over the sides of the body, the gills would lie in a groove between the outer edges of the shield and the sides of the foot (Fig. 5*c*). So long as the animal was relatively slow moving and/or elongately ovate in form, this arrangement would cause little or no difficulties. If an elongated animal moves relatively swiftly on a muddy bottom, however, debris will inevitably be tossed up by its movement. This can potentially foul the gills, interfering with respiration.

There is also another difficulty. Lateral extension of the shell would provide protection for the gills and posterior end of the body. For maximum advantage in locating food or danger, the sensory head should extend at least slightly in front of the shell's anterior edge (Fig. 5*d*). Modern snails withdraw the head into the mantle cavity when threatened with danger, but the "basic mollusk" had no anterior cavity available. Having an "exposed end" that could be attacked without warning would be hazardous to the protomollusk's survival. Extension of the shell plate to protect its backside would have survival value. There would also be the anal opening at its posterior end, and the shell would have to extend at least slightly beyond the anus to allow defecation. This overlap would also result in an opportunity to solve the head protection problem. At this middle stage, the dorsal shield (9) is free on both sides of the body, extends over the anus with a distinct overhang, and also extends as a plate over the head (Fig. 5*c*). It is attached along the center of the body, probably with some muscles that can alter its position slightly.

Respiration would occur as currents of water flowed along the lateral grooves. In time this would lead to the changed gill form seen in Fig. 5*d*, and partial reduction in paired organs, including the attachment muscles. Any reduction in or loss of more anterior muscles would tend toward a situation in which the shell was firmly anchored to the midposterior, but increasingly unattached or loosely attached anteriorly. This would permit a major change that I think led to the molluscan radiation.

If quick muscular action "rocked" the shell forward to extend over the head, this initially would be just as effective as a slight withdrawal of the head under the shell edge. Space under the *posterior* margin of the shell that initially protected the tail and anus would be available to compensate for the *anterior* shell movement. Such forward rocking of the shell would be easier with a domed, limpetlike shell (10) whose anterior portion was longer than the posterior portion (Fig. 5*d*). A slight spiral bend would be more effective than a simple cap-shaped shell, both in terms of rocking action and also because it is lower in total height. Such an animal could have served as direct ancestor to both the Monoplacophora and the Gastropoda. With modifications, it could be altered to give rise to the other molluscan classes. Thus the stage shown in Fig. 5*d* is a critical one. The fact that its shell closely resembles that of the earliest gastropod, *Helcionella,* is deliberate.

The initiation of spiral coiling (12) in mollusks thus could have been an adaptation for forward movement of the shell to protect the head. A by-product of this would have been the expansion of the space at the posterior end of the body under the shell edge. This space would remain freer of turbulence and silt than the lateral margins under the shell edge. For more rapidly moving forms on soft bottoms, concentration of gill filaments nearer the posterior and, eventually, a shift into the posterior cavity would have selective advantage. At the same time, a more rapid reduction in the number of segmental parts could occur, resulting eventually in a paired system of respiratory, excretory, and reproductive organs. Before this point was reached, the cephalopods and monoplacophorans, both of which have forms with four auricles and two or more pairs of gills, would have split off from the basic stock.

A partly segmented organism that had a conical, weakly spiral-coiled shell which could be pulled forward to protect the head (Fig. 5*e*) plus a posterior cavity in which several functions could be concentrated, might have been the basic type of mollusk from which the several extremely successful groups existing today radiated. This last stage has a shell very much like that of the Cambrian gastropod *Coreospira.* While it is definitely "snail like" this still incorporates "protogastropod" features. Did such animals actually exist? Nobody knows. The construction of such hypothetical ancestral forms is one of the great pleasures with which students of animal relationships indulge themselves. The stages shown in Fig. 5 can be added to the long list of proposed "molluscan ancestors." Possibly they will spawn attempts by others to refine details and add new information. If they do, my purpose in presenting them will have been fulfilled.

With this excursion into the molluscan past completed, a survey of the evolutionary trends and diversity within the major classes of living mollusks becomes possible. Chapter IV starts this survey with the chitons and cephalopods.

IV

The Slow and the Quick

Success in our society is frequently equated with becoming conspicuous through the accumulation of wealth, for performance in athletics, or by receiving "star billing" in the entertainment world. Successful politicians, battle winning generals, and Nobel prize winners are recognized for triumphs over rivals, for victories, and for new discoveries.

Our ideas of success involve change, growth, and progress. They are carried over to our view of the living world. Larger size, the presence of more species, and greater activity than their ancestors mean that a group of animals is called "successful" and "progressive."

There is another way to view "success" in the biological world. If organisms have existed with little or no obvious change for over 500,000,000 years, certainly this long survivorship counts as "success." Hardly "modern" or "progressive," but definitely "successful."

Chitons, members of the Class Polyplacophora, first appeared in the fossil record of the Upper Cambrian. While the isolated plates (pieces of the shell) tell us little about the animal, mid-Paleozoic fossil shells differ only in detail from the chitons of today. They were and are slow moving browsers of mostly rocky shores, feeding on algae and small marine animals. With only about 650 living species, chitons cannot be called progressive, diverse, or conspicuous. But much as peasant families continued to till the soil while lordly dynasties came and went, so for more than 500,000,000 years chitons have browsed on littoral sea life as more progressive and larger organisms came and went.

At first glance the cephalopods seem much closer to the usual definition of success. They are active hunters in a variety of marine habitats. Sometimes large and often conspicuous, they have achieved considerable and deserved notoriety (although for the wrong reasons). Cephalopods have somewhere between 650 and 1000 living species, and are therefore only slightly more diverse than the chitons. Although most are small in size, the giant squid, *Architeuthis,* reaches 60 feet in length, including tentacles, while the arm span of a large *Octopus hongkongensis* can reach over 32 feet. Shelled cephalopods today are relatively small. A few *Nautilus* have 10 inch shells and the egg case of the paper nautilus, *Argonauta argo,* can be more than a foot in diameter. Back in the Ordovician, however, there were giant shelled nautiloids, one of which (*Cameroceras*) had a shell nearly 39 feet long, while in the Cretaceous there were coiled ammonoids (*Pachydiscus*) with a shell diameter of about 9½ feet.

During the Ordovician, nautiloids were the largest predators of shallow seas. Today squids share with fish the role of dominant predators in mid-water open ocean areas. More than 10,000 species of fossil cephalopods are known: The "nautiloids" were numerous in the Early and Middle Paleozoic, the "ammonoids" reached their peak in the Late Paleozoic and continued through the Cretaceous, while the coleoids (squids, octopus, cuttlefish) and a few species of *Nautilus* continue to the present day.

Cephalopods are extremely active and conspicuous, and play an important role in ocean ecology. But they are less numerous in terms of species than in past eras, most of their main types are extinct or virtually extinct, and there is little indication that they have major evolutionary potential.

Success is relative and depends on one's point of view. In structure and mode of life, chitons and cephalopods share little in common. Chitons continued the same ways of living from the Cambrian to the present day. Cephalopods experimented first with large shelled predators that were later replaced by sharks and bony fishes. One branch of the cephalopods led to the modern squids, octopuses, and cuttlefish, all of which remain a significant factor in ocean ecology. The slow chitons and quick cephalopods enjoy perhaps equal "success," but in different ways.

CHITONS

Although one family of chitons, the Lepidopleuridae, has been dredged from 15,000 foot depths, most species live in tidal and subtidal areas on rocky shores. Chiton species are particularly numerous along the temperate coasts of Australia and Tasmania (180 species), New Zealand and adjacent islands (71 species), and Western North and Central America (perhaps 165 species). South Africa, Japan, subantarctic shores, and the Caribbean region are secondary centers of diversification.

Chitons are sometimes extremely abundant. In the 1920s, before the great increase in harbor pollution, Tom Iredale could collect during a single low tide in Sydney, Australia at least 1000 specimens of more than 20 species. In such a situation, a few species will live on the upper side of rocks that are exposed during low tides, most will hide on the undersurfaces of rocks that are only briefly or rarely exposed at low tide, and others will cling to various seaweeds.

How can so many kinds of chitons live in one place? It is a basic biological law that "no two species can live in the same way at the same place at the same time." This idea of "competitive exclusion" or "Gause's principle" (the Russian scientist Gause first demonstrated this experimentally) explains why very closely related species do not live in the same area and why the species that do live in the same tide pool (or forest) must do so in slightly different ways. These 20 species of chitons in Sydney Harbor each show a different combination of preferences for shelter site, position in the tidal zone, food, activity period, and/or many other factors. While they share the same *habitat,* in this case a tide pool. their *niche,* the role they play in the tide pool community, is different for each species. Niche is a term used by ecologists to indicate not only the physical place occupied by a species (under big rocks just below the low tide level as opposed to the upper surface of rocks near the high tide mark), but where it fits into the "eat or be eaten" chain, and what its preferences are in relation to such factors as light, temperature, humidity, salinity, surface features, and other organisms not involved in their particular "eat or be eaten" network.

Manuals describing local faunas or guidebooks to seashore life are concerned with identifying species, describing the different habitats, and then pointing out niche factors to enable seeing or collecting particular species. However, for most mollusks the exact niche requirements remain unknown. Generations of naturalists will be required to observe and record sufficient data so that the habits of mollusks will be as well known as those of birds. For no area of the world is the chiton fauna as a whole well enough known so that we can describe the niche differences between species. In some areas the habitats can be described and the chiton's physical position in a tide pool pinpointed, but we have only vague ideas as to their exact roles.

Seen from above on the edge of a tide pool, a chiton is often not very attractive, although a few are brightly colored (Plate 2, center right). The shell surface is frequently eroded, or covered with barnacles, moss animals, algae, or other mollusks. Up to 125 barnacles or more than 25 different kinds of plants and animals have been found on a single chiton in the mid or upper tidal area. Chitons from the underside of rocks that are lower in the tidal zone often will be far less encumbered by other organisms. Their structure is much easier to observe.

Most obvious is the leathery *girdle* that borders the body and holds together the eight partly overlapping shell *plates* (Fig. 1). The girdle is a muscular tissue that is highly flexible; its edges can be molded exactly to the uneven surface of a rock. Any point along its margin can be lifted so that water for respiration can enter, pass posteriorly, and then exit near the posterior end, perhaps accompanied by undigested food or excreted matter. The surfaces of both the girdle and shell plates are variously "decorated" with scales or other projections that can be calcareous or chitinous. Often the combined nature of the girdle and shell plate sculpture will provide the quickest way to identify local species. Examples of typical sculpture are shown on the shell plates and along the girdle outline (Fig. 1). The width of the girdle and the extent to which it partly overlaps the shell plates vary greatly. In the Australian *Cryptoplax* and *Acanthochiton*, the girdle covers much of the plates, whereas in *Lepidopleurus* it is only a narrow fringe (Fig. 2). The giant *Cryptochiton stelleri* of Western North America, which reaches 14 inches in length, has the plates completely covered by the girdle.

The shell plates are an *anterior* or "head" plate, *posterior* or "tail" plate, and six intermediate plates. Seen from outside (Fig. 3*a*) the intermediate plates show a central region divided into two *pleural* (2) and one *jugal* (1) section, plus two *lateral* (3) areas, which often have different types of sculpture. Projecting anteriorly from the inner part of the shell are two *sutural laminae* (4) that flank a *jugal sinus* (5). These laminae extend under the posterior margin of the next anterior valve. This overlapping (see Fig. 1, Chapter III) permits a chiton to roll into an armored ball like a sow bug or armadillo. Not all chitons have these sutural laminae. They are absent in the earliest fossil chitons and in living groups that are considered primitive. Particularly prominent on the outer sides (Fig. 3*b*) of the anterior and posterior plates are the *insertion teeth* (6), which unite the girdle with the plates. Species that live on exposed rocks pounded by waves have much stronger insertion teeth than do chitons that live in more protected sites. Similar, but usually weaker, insertion teeth are visible on the outer edges of the intermediate plates. In many species these plates show marked *slit rays* (7), which are channels for nerves to the outer body surface. Each plate is composed of complex structural layers which are grouped into two categories: an outer, highly porous, softer group of tissues forming the *teg-*

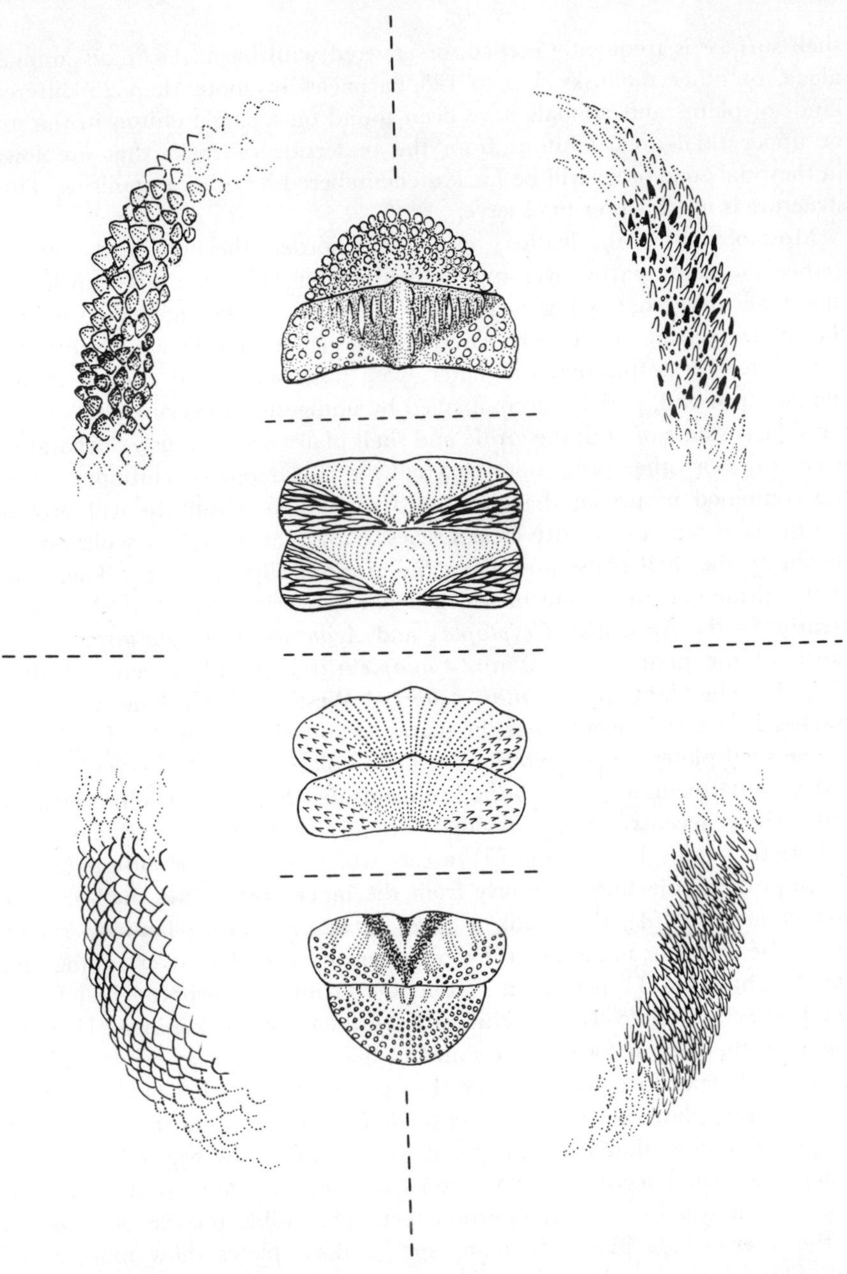

FIGURE 1. Chiton showing variations in girdle and shell plate sculpture. Modified from various sources and specimens.

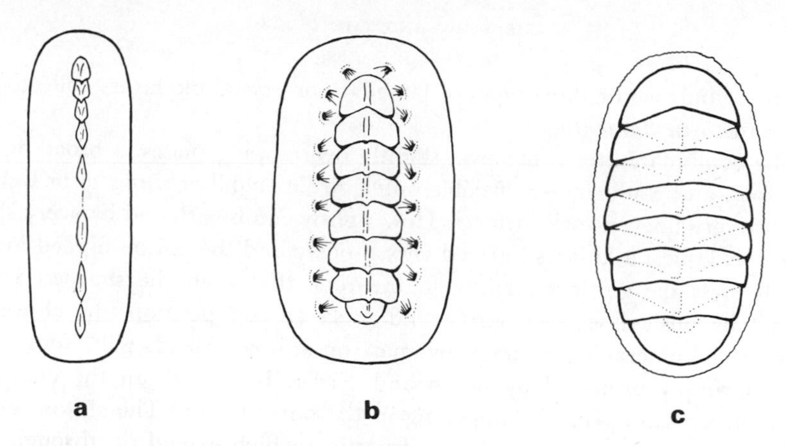

FIGURE 2. Extent of dorsal girdle overlap in: (*a*) *Cryptoplax* from North Queensland, Australia; (*b*) *Acanthochiton* from Victoria, Australia; and (*c*) *Lepidopleurus* from New South Wales, Australia. The bristles in *b* mark the lateral edges of the plates and show the amount of overlap. Adapted from Iredale and Hull.

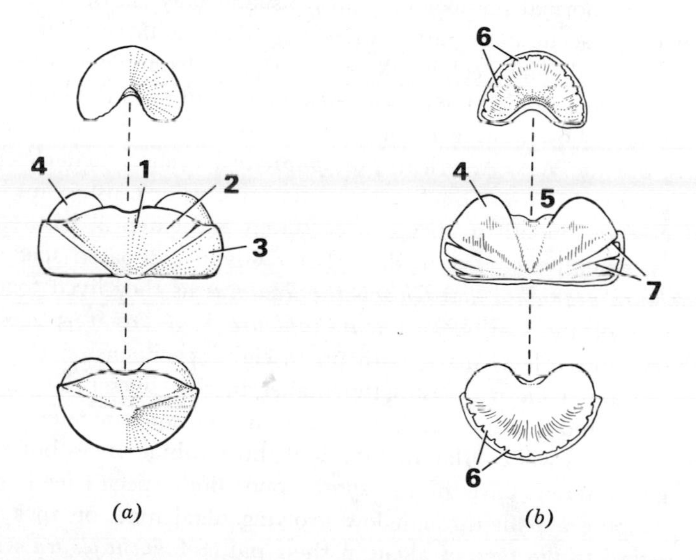

FIGURE 3. Outside (*a*) and inside (*b*) views of the anterior (top), intermediate (middle), and posterior (bottom) plates of the West American *Ischnochiton conspicuus*. Structures shown are jugal (1), pleural (2), and lateral (3) sections of the intermediate plate, sutural laminae (4), jugal sinus (5), insertion teeth (6), and slit rays (7).

mentum, and two or three types of lamellate or crystalline layers collectively called the *articulamentum.*

The combination of protective, slightly overlapping plates, a broad adhesive foot, and a tough, very flexible outer girdle enables chitons to fit tightly against very uneven rock surfaces. How tightly can be attested by every shell collector. Unless a knife is inserted very quickly and the chiton flipped loose, it will press the girdle so tightly to the rock that even the sharpest knife cannot be pushed between girdle and rock. In this position, the chiton is virtually invulnerable to attack by man or animal. Head, gills, foot, and tail are amply protected by plates and girdle. But how can the clamped down chiton sense what is happening in the outer world? The chiton senses by means of numerous microscopic structures which extend up through the shell layers. In the Chitonidae, for example, the anterior plate alone may have 500 to 3000 microscopic "eyes," while pores of smaller size act as chemoreceptors or tactile organs. The plate surfaces of chitons have scattered sensors instead of sensors which are concentrated in the head region, as in snails, cephalopods, or other more active animals.

Repaired shell injuries are common in chitons because of exposure to battering waves and tumbling rocks. Perhaps one in every 500 chiton specimens will have an abnormal number of plates, usually only six or seven, caused by fusion of adjacent plates after an injury. Even a three plate specimen has been recorded, and occasionally a nine plate example is seen.

Few people have studied the ecology of chitons, but even the limited observations to date show a wide range of behavior. Most browse on algae or bryozoans, but the West American *Placiphorella* has short tentacles projecting from the head that are used to trap small worms or crustaceans between the tentacles and mouth so they can be seized and eaten at leisure. In one of the most extensive studies to date, Peter Glynn compared specimens of *Acanthopleura granulata* and *Chiton tuberculatus* as they lived together in several areas of the Caribbean. *Acanthopleura* is a larger species, living above the mean sea level mark, with up to eight specimens on each square meter of surface. *Chiton* is distinctly smaller in size, lives below the mean sea level mark, and has up to 22 individuals on each square meter of surface. Their distributions will overlap in coral reef shore rubble areas, but generally they occupy different parts of the shore zone. Both species feed primarily at night, rasping a path through low growing algal mats on rock surfaces without regard to the type of algae in their path. *Acanthopleura* will move and feed at night even when exposed by low tides, averaging 8 hours of feeding each night, while *Chiton* feeds an average of only 6 or 7 hours a night. Both species grow quite rapidly, reaching a size of about 1½ inches in their first year of life. Differences in the timing and amount of reproductive activity relate to their habitat, with the more frequently exposed *Acanthopleura* spawning at more precise intervals and in smaller quantity than does *Chiton,* which is seldom exposed to the air and thus can count on its spawn being released into water virtually all the time. Few animals are known to predate on the larger specimens of these chitons; kills from exposure by exceptionally low tides, or when severe storms reduce the amount of rubble under which they can hide are major factors in controlling their individual abundances in a given area.

Chitons are tough-coated, persistent, browsing inhabitants of rocky shores. Their absence from sandy or muddy bottoms may be connected with silt fouling their gills or the vulnerability of their underside. Their success is the result of their perfect adaptation to the rocky shore wave zone.

CEPHALOPODS

The development of tentacles for catching and holding prey, formation of a jet propulsion device, plus vastly improved nervous and sensory systems are the keys to cephalopod success. Nearly all modern species are free from the constrictions of a shell, but during their zenith in Paleozoic and Mesozoic seas, their shells were of fantastic size and shape (Figs. 5 and 6). Rapid, darting movements and presence of a heavy shell are contradictory. It has been difficult to imagine exactly how the process of altering a basic mollusk to the cephalopod type started, but we can identify some of the key changes. New ideas have come recently from studies of fossils, but many have come from studying the only living shelled cephalopods, members of the genus *Nautilus* (Plate 1). *Nautilus* is a living relict of past ages. It is found in shallow to moderately deep waters in the Southwest Pacific between the northern Philippines and the Fiji Islands, although there are also a few populations living near the southern coast of Australia.

Seen in swimming position (Plate 1, top), the two small ocular tentacles just above the slitted eye, and part of the outer circle of digital tentacles are obvious. The more numerous inner circle of labial tentacles is hidden in this view. Projecting from a few of the lower tentacles are tips of the retractile *cirrus,* which is a prehensile and adhesive grasping organ. True suckers of the type found in squids and octopuses are absent. It is the cirrus (Plate 1, bottom) that grabs and holds prey or other objects; it is capable of very great extension, and is very flexible. Some of the tentacles are tactile, while others are olfactory, since *Nautilus* depends more on smell and touch to locate food than on its large but weak eyes. Near the lower edge of the shell, the circular funnel or *hyponome* projects. This fold of the mantle can be turned in many directions, with expelled water moving *Nautilus* in opposite direction from the jet stream. Water enters the sides of the mantle cavity, sweeps across the four gills, and passes the digestive and reproductive system openings before muscular contractions force it out of the hyponome. The roughly triangular upper part of the projecting animal is the shieldlike *hood,* a red and white muscular lobe that exactly fits the shell opening. Both hyponome and tentacles can be withdrawn into the shell and the hood pulled down either to block the opening completely or to leave only the eyes exposed.

Observations on living *Nautilus* by Anna Bidder and others show that during the day *Nautilus* rests on or near the bottom, moving about at night in search of food. The ocular and varying number of the digital tentacles are spread widely into a "cone of search" when the animal is either hunting food or exploring a new area. When a food item is touched, the cirri on the inner digital tentacles are protruded to grasp the prey and pull it toward the mouth, where a powerful beak and the radula tear it to pieces. The bottom figure of Plate 1 clearly shows the exploratory use of the outer tentacles. The upper photograph shows only partial extrusion into a "cone of search," since the animal is swimming away from the photographer.

The shell of *Nautilus* consists of about three whorls that rapidly increase in size. If sawed in half (Fig. 4), one of the key cephalopod adaptations is revealed. Most of the upper shell is divided into a series of chambers by curved *septa.* These can number 33 to 36 in adult examples. The living animal occupies only the outermost area of the shell, with the many inner chambers filled with air and a small quantity of fluid. In life, a part of the animal called the *siphuncle* forms a continuous tube extending from the animal through to the uppermost chamber of the shell. In most dried museum specimens, such as that shown in Fig. 4, the connecting links of the siphuncle are broken. The chambers are filled with gas and serve as a flotation mechanism. Buoyancy provided by the gas in the chambers counterbalances the weight of the shell and animal, which can float, slowly rise, or slowly sink by minutely adjusting the ratio of fluid to air within the inner chambers. Water jetted from the hyponome is used for directional movement and is not required to keep the animal from sinking.

This means of floating the shell, coupled with the formation of prehensile tentacles and the jet propulsion system are the basic changes that led to cephalopod successes. The hyponome is a modified part of the molluscan foot, but the tentacles are probably a modification of head structures rather than the foot, as has been claimed.

Evidence from early fossils led E. L. Yochelson, R. H. Flower, and G. F. Webers in 1973 to propose that cephalopods were derived from the

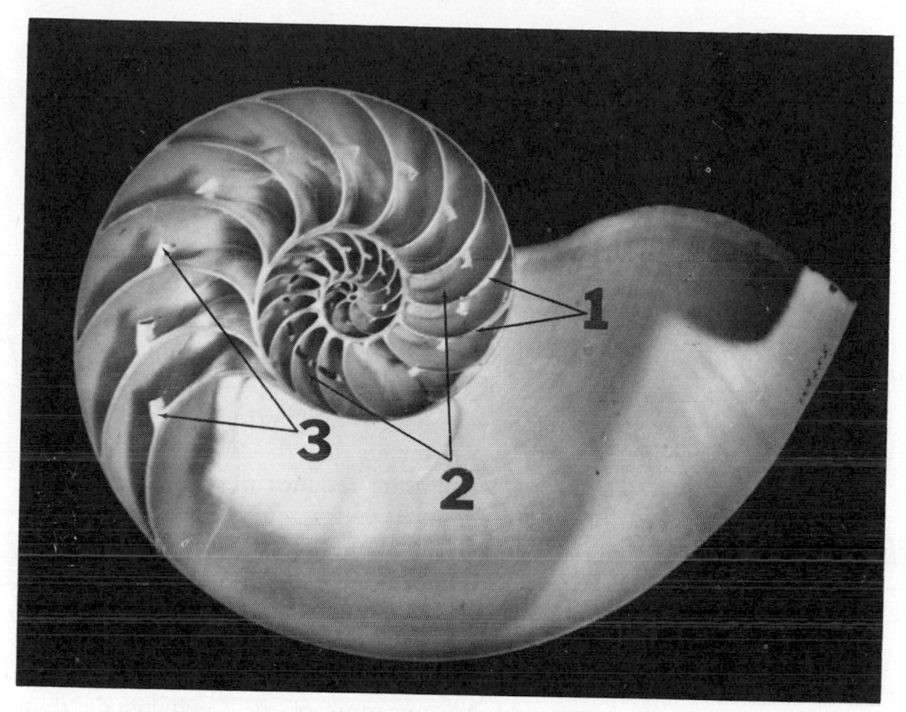

FIGURE 4. Internal view of a *Nautilus* shell that has been cut in half, showing septa (1), chambers (2), and remnants of the siphuncle (3).

Monoplacophora. Several Late Cambrian monoplacophorans and gastropods show septa dividing off parts of the upper shell. While none of these showed any indication of a siphuncle or hole in the septa, the presence of experiments in partitioning the shell during the Late Cambrian is well documented. The change from these high-spired, septate conical shells to the first known cephalopod, *Plectronoceras* from the Late Cambrian of Manchuria and North China, involves mainly sinuation of the outer lip and the all important formation of a siphuncle to permit using the chambers between the septa as a buoyancy device. The authors speculate that the "presiphuncle" could have originated as a strand of body tissue that failed to be withdrawn during final deposition of the first septum, assuming quite logically that the septa must be secreted from shell sides toward the center. At first this strand may have had no function, but subsequently became modified to act as in *Nautilus*. Whether liquid was secreted at first deliberately into the chambers or was accidentally introduced during chamber formation is unknown. Early cephalopods are thought to include both buoyant swimming forms which held the shell in a vertical position, such as *Plectronoceras,* and bottom dwellers in which the shell was horizontal. The former would have had mostly gas in the upper chambers, the latter a higher proportion of liquid. In time, the ability to adjust the proportions of gas and liquid would have evolved and permitted the great expansion and experimentation of the Cephalopoda in the Paleozoic and Mesozoic.

The theory just described, if correct, brings the origins of the Gastropoda, Monoplacophora, and Cephalopoda converging toward the hypothetical mollusk proposed in Chapter III. It would mean that the cephalopods arose through partitioning of a high conical shell followed by accidental development of buoyancy as a by-product This permitted exploitation of a vastly different adaptive zone from that pursued by the gastropods (see Chapter VI), whose asymmetrical coiling and torsion led to effective hunting on or in the sea bottom. The monoplacophorans retained the primitive caplike shell and did not radiate into any major adaptive zone, at least during the Lower and Middle Cambrian. Elongation of their shell led to septal formation during the Late Cambrian, by which time the gastropod radiation was well under way. Gastropods also show septate forms, and indeed many living taxa use septa or apical plugs to wall off early whorls as growth proceeds, but these are parallel changes. The monoplacophorans, with their partial segmentation, stand as more probable ancestors of the Cephalopoda than the Gastropoda, since there are cephalopods with four auricles and two or more gill pairs.

While C. M. Yonge and his students maintain that mollusks were never segmented, I agree with Lemche and others that mollusks were originally at least partly segmented. These theories concerning the origin of mollusks and the different main groups are controversial. All such ideas, to quote Yochelson, Flower, and Webers only slightly out of context, are ". . . built mainly of bricks of speculation held together loosely by a thin mortar of facts." Nevertheless, the ideas developed here on the origin of the molluscan shell, spiral coiling, and torsion mesh with the theories of Yochelson, Flower, and Webers to provide a logical and coherent thesis concerning molluscan evolution. Other viewpoints are expressed in the books listed in Appendix B. But this digresses from the cephalopods.

Traditionally the fossil cephalopods have been placed in two subclasses: the Nautiloidea, which includes those species with relatively simple septa and shell sculpture such as the living *Nautilus,* and the Ammonoidea for those species with complex septa and often very elaborate surface sculpture. More recent studies demonstrate that recognition of five primarily fossil subclasses in addition to the living Coleoidea (squids, octopuses, and cuttlefish) gives a more balanced picture of relationships and evolution. Nevertheless, the terms "nautiloid" and "ammonoid" are useful for shell types.

Perhaps the largest invertebrates that ever lived were cephalopods of the Ordovician Period. Reconstructions of the apparently bottom dwelling *Endoceras* and a school of the free swimming *Ormoceras* (Fig. 5) indicate two nautiloid shell types of Paleozoic seas. By the late Cretaceous, the "ammonoid" types had replaced the nautiloids, producing (Fig. 6) such odd looking creatures as the scalariform and spinose *Hamites navarroense* sitting just to the left of the evenly coiled *Placenticeras placenta*. The latter reached almost 2 feet in diameter. A school of *Baculites ovatus,* a straight-shelled form with minute spiral tip, hovers above the sea bottom. With the exception of the few *Nautilus,* ammonoid and nautiloid stocks alike became extinct, while another group probably evolved into the modern squids and octopuses. Appearing in the Mississippian portion of the Carboniferous, belemnite cephalopods had an internal chambered shell and perhaps were ancestral to the modern coleoids. By the late Jurassic, quite squidlike belemnites such

FIGURE 5. Reconstruction of life in a Late Ordovician sea, including the bottom dwelling nautiloid cephalopod *Endoceras* and a school of the free swimming nautiloid *Ormoceras.* A trilobite (extinct Paleozoic arthropod) is above the head of *Endoceras,* and an early snail is in the immediate foreground. Courtesy Field Museum of Natural History, Chicago.

FIGURE 6. Reconstruction of Late Cretaceous sea life, including the large spiral ammonoid *Placenticeras placenta,* a school of *Baculites ovatus,* and the "cork screw"-shaped *Hamites navarroense.* Several marine snails are in the foreground. Courtesy Field Museum of Natural History, Chicago.

as *Pachyteuthis* were schooling in the oceans (Fig. 7). All of these reconstructions are based on the fossils having the same tentacles and head structures seen in modern cephalopods. We have no direct knowledge of the bodies belonging to the fossil nautiloids or ammonoids, so their tentacles may have been quite different.

Cephalopods were the first large swimming predators of ancient seas. They were joined by the arthropods (scorpionlike eurypterids) and vertebrates

FIGURE 7. Reconstruction of Late Jurassic belemnite cephalopods of the genus *Pachyteuthis* hovering over a bed of the oyster *Gryphaea nebrascensis*. Courtesy Field Museum of Natural History, Chicago.

(fishlike ostracoderms) in the Late Ordovician and Silurian. Through the rest of the Paleozoic, cephalopods shared the role of primary oceanic predators with the sharklike fishes. In the Early Mesozoic various fish groups and the huge marine reptiles (ichthyosaurs, plesiosaurs, and pliosaurs) successfully competed with the shelled cephalopods. Eventually the modern bony fish groups, sharks, and a new group of cephalopods, the coleoids, came into a competitive balance. A. Packard in 1972 presented a very stimulating review of the ways in which cephalopods and fish converged during evolution. He hypothesized that at first the early teleost fishes excluded the cephalopods from the surface and coastal waters, with the subsequent invasion of these areas by modern octopods and cuttlefish representing a renewal of successful competition between fish and cephalopods by the latter. The basic behavioral and growth differences between fish and cephalopods are the cephalopods having a locomotion pattern that permits moving forward to attack, then backward to retreat, usually parental care or special handling of the eggs, growth to maturity in only 1 or 2 years, and marked daily vertical migrations accompanied by activity during either dawn or twilight hours. The latter feature was pointed out by Packard to characterize less advanced fish groups also. The modern cephalopods are most abundant in midwaters and at depths of more than 1000 meters, whereas modern fishes are more diverse in coastal waters.

Living coleoids belong to three general types—octopuses, squids, and cuttlefish. In general, octopuses are eight-armed inhabitants of rocky shores or reefs, lack any internal shell, and tend toward occupying a permanent lair, which can be either a natural hole, a pile of rubble constructed by the animal, or in the case of a small Caribbean species, a dead clam shell whose valves are still connected by the ligament is appropriated. The octopus retreats inside, and then uses its suckers to close the two valves together, providing protection equal to that enjoyed by its original owner. Many octopod species have proved to be excellent experimental animals, quite amenable to training, and with superb sensory equipment. Capable of rapidly changing color to match backgrounds or to signal moods, octopuses are among the most interesting shore area organisms. Movement is by slithering along on the arms or jetting through the water with the muscular closed funnel. A few are known to have a poisonous bite, particularly a brilliantly blue-ringed Australian species, *Hapalochlaena maculosa,* but most are harmless to man. Among the oddities of the Octopoda are the small deep water, floating species, *Cirrothauma,* which has webbing between the eight arms, and the large pelagic *Argonauta.* The females of *Argonauta* have a greatly enlarged, paddle-shaped first pair of arms which secrete a thin shell-like structure (Fig. 8) that is then held around the body and also used to carry eggs. Large, unbroken specimens of this "shell" are greatly prized by collectors.

Squids and cuttlefish differ most obviously from octopods in having an internal supporting skeleton and a new pair of extensible arms that are much longer than the eight basic arms. The ninth and tenth arms are used to capture prey. Cuttlefish (Fig. 9) live in shallow water, and often burrow just beneath the surface of the sand or dart about in search of food. Their large internal skeleton is the "cuttlefish bone" placed in cages for canaries. This skeleton is the buoyancy device for the cuttlefish. The porous skeleton is filled partly with gas and partly with fluid, the animal adjusting the ratio to maintain the desired position. A fin encircles the body (Fig. 9) and is used to balance the animal, who progresses by jets of water from the funnel.

Squids have a much smaller skeleton that does not function as a buoyancy mechanism. Instead the squid must constantly use water jets to maintain its position. The posterior fins (Fig. 10) act as balancing mechanisms. Squids

FIGURE 8. Shell of *Argonauta argo*, a pelagic octopod. The specimen is $7\frac{1}{4}$ inches long.

are inhabitants of the open waters, although species such as *Loligo* frequently occur near shore. In terms of volume squids are among the largest components of midwater life in the open sea. Because of their acute senses and rapid movements, they are hard to catch in nets or trawls. As a result, we know little about the diversity of squids from the open oceans. Sperm whales feed exclusively on squids. Calculations as to the number of squids needed each year to feed the estimated number of living sperm whales suggest that an awesome number of squids are swimming about. One sperm whale stomach contained 2136 cephalopod beaks, while the annual catch of a common Japanese squid, *Todarodes pacificus*, exceeds the total catch of herring in the entire Atlantic Ocean.

Coleoids have tentacles equipped with highly efficient suckers. These are cups which range from a fraction of an inch up to $2\frac{1}{2}$ inches in diameter, and often have the rim lined with a row of teeth or claws. When a muscular stalk that leads to the body contracts, the cup is enlarged, creating a partial vacuum that locks the cup to its prey. Relaxation of this muscle releases

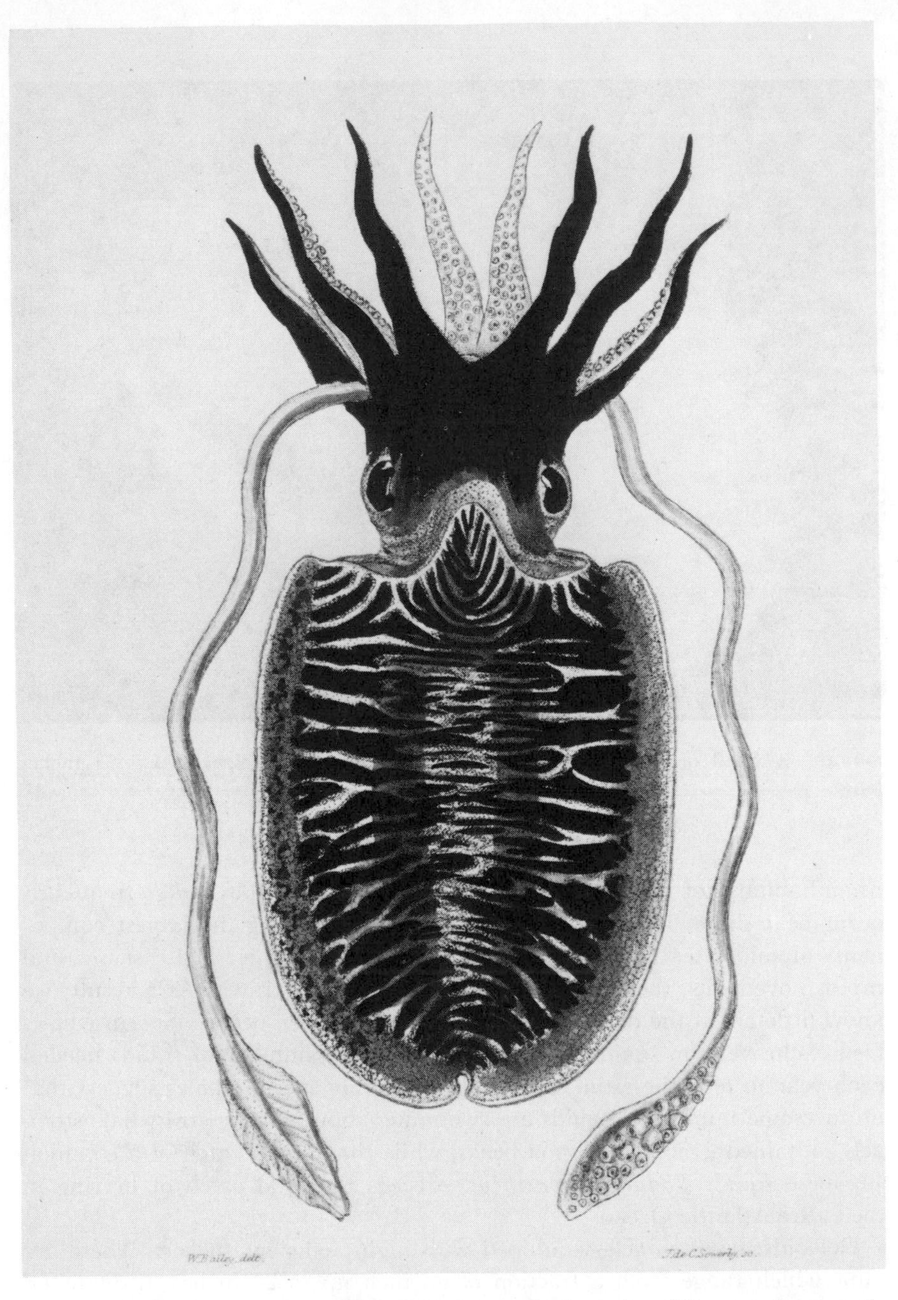

FIGURE 9. Cuttlefish, *Sepia officinalis,* drawn to show the difference in tentacle length and sucker patterns. From Forbes and Hanley.

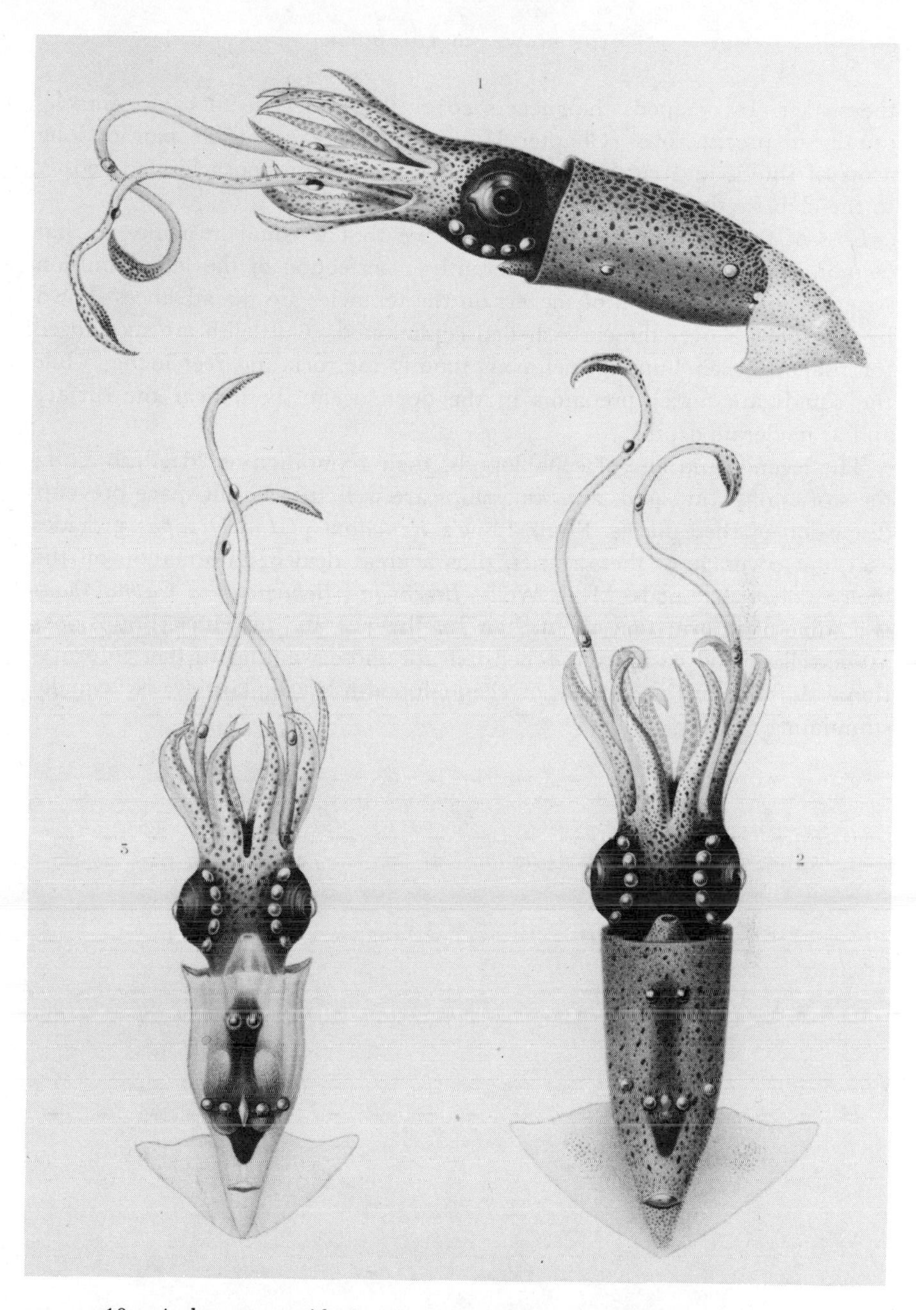

FIGURE 10. A deep-sea squid, *Lycoteuthis diadema,* collected in the South Atlantic Ocean at 5000 to 10,000 foot depths. Rows of beadlike light-producing organs and the enormous eyes identify this as a deep-sea animal. These structures are absent in shallow water species. Total length of animal including tentacles is about 3 inches. From Chun.

the sucker. In octopods the suckers cover the underside of each tentacle, usually in precise rows (Chapter II, Fig. 8). The extensible pair of long arms of squids and cuttlefish have either suckers or hooks limited mostly to the clublike tip of the arms.

Loss of the shell, development of an eye that is equal in acuity to that of man (although functioning differently), perfection of the jet propulsion system, and the evolution of suckers on the tentacles are the advances shown by the coleoids over the early shelled cephalopods. Cuttlefish are specialized for shallow water hunting, octopods mainly for rock and reef living, while the squids are major predators in the open ocean, both near the surface and at moderate depths.

The legends and lore of cephalopods, their reputation as "devilfish," and the stories of giant squids attacking ships are well known, but space prevents discussion of these facets. Frank Lane's *Kingdom of the Octopus* provides excellent coverage of these stories, plus a great deal of information on the biology of cephalopods. M. J. Wells' *Brain and Behaviour in Cephalopods* is a mine of information about their intelligence and activities. Both books are excellent and easily comprehensible for those wanting further information. A. Packard's review of cephalopod-fish similarities is equally stimulating.

V

Filtrationists

After a storm the beaches of West Florida are covered with dead and dying mollusks being picked over by a hoard of shell collectors. There are perhaps 20 or 30 clams for every snail shell. This is not an accident. Dredged samples from the bottom of Lake Michigan, the North Sea, and ocean trenches alike show that clams are the most abundant mollusks on soft bottoms. A large collection of fossils from Newport Bay, California, which was studied by George Kanakoff and William Emerson, contained 81,301 shells. The 128 bivalve species averaged 303 specimens each. There were many more kinds of snails, 289 species, but these averaged only 136 specimens per species.

Since clams generally sit quietly and pump water in and out of their bodies, or at most roam slowly through the bottom ooze, their great abundance seems paradoxical. It relates to the fact that food production on land and in the sea is quite different. The base of the land food chain (producers) is the green plants, from lichens and mosses to grasses, shrubs, and trees. Animal life must move about to feed on the stationary growing plants, or hunt other animals. Undigested food and excretory products from animals are dropped on the ground and become available as food for other organisms.

In the ocean, the base of the food chain lies with the microscopic acellular diatoms, dinoflagellates, microflagellates, and other types of algae that float in the upper layers of the water. The numbers of individuals are astronomical. Feeding on these are innumerable larvae of marine animals, small crustaceans of various kinds, and a few mollusks such as the heteropods and pteropods. These in turn are fed upon by larger crustaceans, arrow worms, jellyfish, and small fish, who in turn are eaten by larger fish, sea birds, marine mammals, and squids. Undigested food and excretory products from all these organisms are released into the water, where they either dissolve or sink slowly downward. Either they are eaten on the way down, or in time, large quantities of these accumulate on the sea bottom as deposits of organic matter and chemicals. In very productive ocean waters, the range of life from diatoms to small crustaceans in sufficient water to fill a large classroom (30 feet square with 10 foot ceilings) could fit easily into a pint jar. These organisms are scattered through the upper 100 to 300 feet of the open ocean. Not only is the amount of organic matter produced each year in an area of open ocean usually much less than that produced in an equivalent area of productive land, but the food on land is concentrated within either a few inches below or a few feet above the soil surface. Plant life on land is large, stationary, obvious, and within a limited vertical space.

Harvesting food producers in the sea requires extracting numerous very tiny organisms from a huge volume of water. The larval hunters of the plankton, and the crustaceans or small fish that feed on them in turn, dart and move about while hunting their tiny prey. Other animals, such as the coelenterates, float or are attached, using their stinging-cell equipped tentacles to grasp and poison prey that floats or swims within their reach. Tube worms and barnacles jerk their "arms" in and out, creating currents of water and grabbing food items. Brachiopods and bryozoans use their ciliated "lophophore" to entrap the microscopic particles or even organic molecules on which they feed. Clams pump a current of water into their body, use their gills to sort out diatoms and other small organisms, and then return the water to the exterior. Clams are filtrationists *par excellence,* exploiting food organisms and particles in the water (suspension feeders) and those accumulated in debris lying on the bottom (deposit feeders). Marine and brackish water species are surveyed here, and the freshwater groups are discussed in Chapter VIII.

Because they have adapted to such diverse habitats as boring in wood or rock, fastening by fibers or cementing the shell to a fixed surface, burrowing in the bottom silt, or a few that swim by opening and closing their valves, an initial impression of bivalves concerns their variety in shape, size, color, and shell sculpture (Plate 2). The few spines on the burrowing *Pitar* (upper left), the touch of color between the curved thicket of spines on *Spondylus* (lower right), the contrast in shape and color between the anterior and posterior ends of *Corculum* (upper right), and the variable color of *Chlamys* (Plate 3, upper left), with its blue-eyed mantle (upper right), give only a faint indication of bivalve diversity. This diversity, which relates to the mode of living, masks what is basically a highly conservative body plan. The cluster of thin *Isognomon alata* attached to a rock bear little resemblance to the large, white, delicately sculptured shell of *Barnea costata* (Fig. 1), although both live in the same bay off West Florida.

FIGURE 1. Shell of *Barnea costata* (single white shell) and a group of *Isognomon alata* attached to a rock fragment. Courtesy Field Museum of Natural History, Chicago.

Inspection of an empty clam shell, such as *Mercenaria campechiensis* (Fig. 2), shows that it is a highly integrated unit. On its upper or *dorsal*

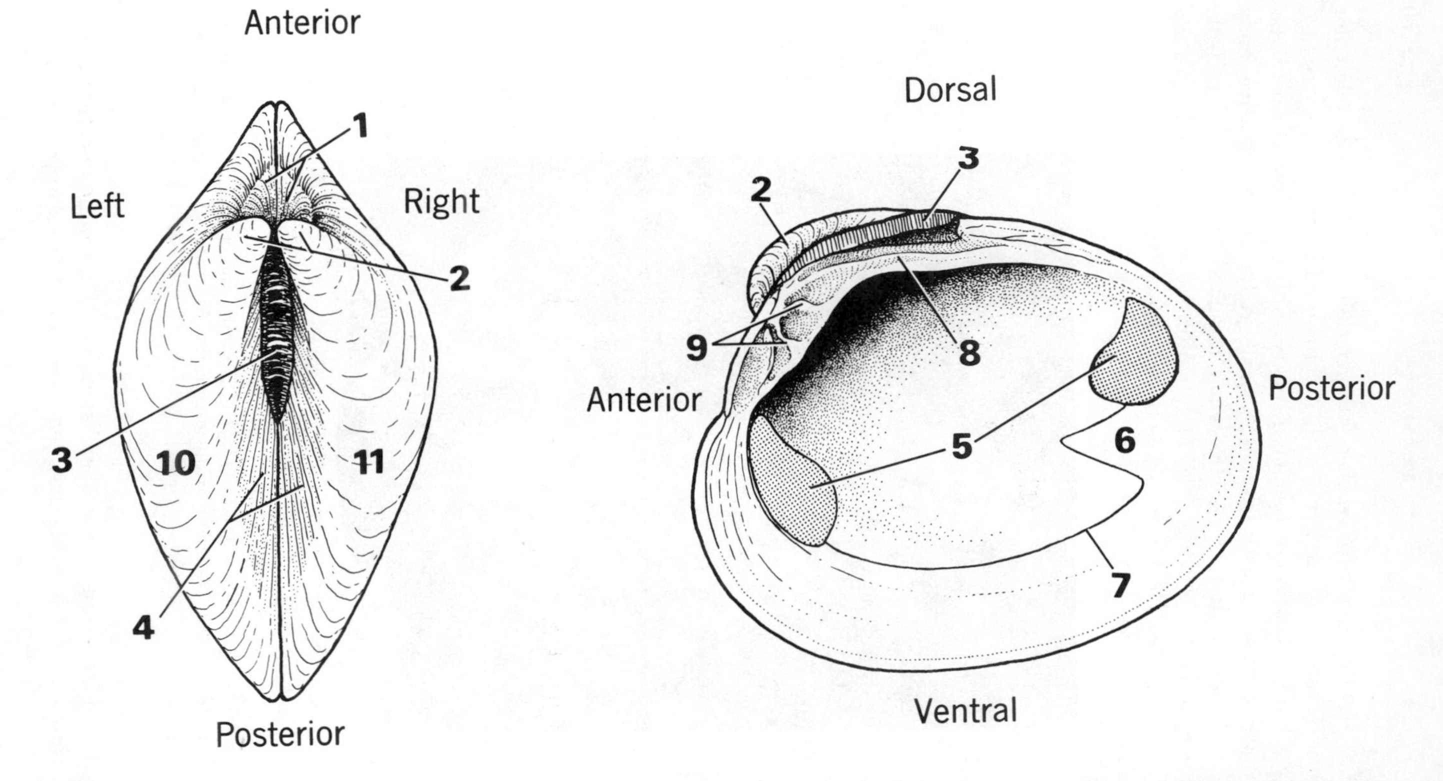

FIGURE 2. Dorsal (left) and internal (right) views of a clam shell (*Mercenaria campechiensis*). Structures shown are lunule (1), umbos (2), ligament (3), escutcheon (4), adductor muscle scars (5), pallial sinus (6), pallial line (7), lateral tooth (8), cardinal teeth (9), left valve (10), and right valve (11).

side are the pair of *umbos* (2) or beaks. They are the first calcified parts of the shell, which grows by a series of concentric additions to the outer margins. Usually just posterior to the umbos is a flexible organic structure, the *ligament* (3), which is composed of several different layers. This structure is normally the axis on which the shell pivots in opening and closing. The ligament is flexible and functions to keep the two lateral calcified parts of the shell, the *valves* (10, 11), gaping slightly to moderately open. In typical bivalves the ligament lies above the hinge line, but in many species it has shifted to an internal position (Figs. 4*c, d*) between the hinge teeth. In this position it can prop open the shell without causing the teeth to unlock.

On either side of the ligament and often extending further posteriorly are the incurved *escutcheons* (4) of each valve, while just anterior to the umbos is the depression called the *lunule* (1). Frequently the sculpture and structural details of the umbos, lunule, escutcheon, and ligament are of great help in identifying bivalve species. If the shell is held so that the umbos and ligament are as in the left diagram of Fig. 2, with the ligament nearest to you, then the left valve will be on your left and the right valve on your right.

Valves and ligament are a single developmental unit, probably derived by decalcifying the median ridge of a high, narrow conical shell and replacing this ridge with the resilient and flexible ligament. The earliest unquestioned bivalves are Late Cambrian in age, but there are several problematic fossil groups that may or may not be related. In 1973, restudy of the problematic Early Cambrian organism *Fordilla troyensis* led several leading paleontologists to conclude that it was a clam, not a bivalved crustacean. An early evolutionary experiment in or from the bivalve direction is the Ordovician to Late Permian group Rostroconchia, in which a cap-shaped larval shell sits between the umbos. Rostroconchians lack any trace of a ligament and apparently had the shell fused dorsally. Exactly where this group fits into the bivalve lineage is still uncertain. Some scientists have suggested that it be isolated as a separate class of mollusks, but its conformance to the basic bivalve condition is much greater than the cited differences. The Rostroconchia and Stenothecoida (another obscure fossil group) seem better classified as aberrant bivalve experiments, rather than as distinct classes of mollusks. Yet other unknowns are the fossils assigned to the classes Hyolitha and Mattheva, extinct animals that may not be molluscan at all.

Looking at the inside of a right valve (right diagram of Fig. 2), the most obvious dorsal structures are the *hinge teeth* (8 and 9), located below the broken ligament. The teeth function to guide the two valves together in such a way that the outer edges meet exactly when the shell is closed. They also prevent the valves from shifting anteriorly or posteriorly in relation to each other. Bivalves that are exposed to pounding waves, surf, or strong predators tend to have very massive and positively interlocking teeth (Figs. 3, 4*e*), whereas species that burrow deep in the bottom muck or lie in rock holes tend to have greatly reduced teeth. Those teeth lying below the umbos are called *cardinal* (9) teeth, and those further removed from the umbos are called *lateral* (8) teeth. A projecting tooth on the hinge of one valve will be matched by a corresponding socket in the opposite valve.

FIGURE 3. Shell of *Spondylus americanus* opened to show hinge interlock and pattern of surface spines.

Slightly above the middle and near each end of the valve are the impressions of the *adductor muscles* (5). These often deeply incised pits are the points of attachment for the transverse muscles that close and hold the shell shut. A quick-acting set of fibers acts to pull the valves together, while less responsive but more powerful "catch" muscle fibers hold the valves shut against the resiliency of the ligament which tends to gape the shells open. Often extending from the anterior adductor muscle scar to the posterior adductor muscle scar is a sinuated line that marks another set of muscle attachments. The lower *pallial line* (7) is where the mantle is fastened to the shell by the pallial muscles. The *pallial sinus* (6), when present, marks the insertion of the muscles used to retract the siphons into the shell. If the siphons are not retractable or very short, then the pallial sinus will be absent. In many of the actively burrowing clams, additional attachment scars on the shell show where the foot retractor muscles fasten.

The pattern of muscle scars on the shell imparts information about the structure of the clam, including whether the siphons are retractable and something of their relative size, an indication of the foot size and its degree of use in burrowing, and the pattern of shell closure. At one time the relative size and number of adductor muscles were used to recognize orders of clams, but subsequent studies have downgraded their importance. If the posterior and anterior adductors are both present, the clams are called *dimyarian;* if they are equal or nearly equal in size, the clams are called *isomyarian;* if the anterior adductor is greatly reduced in size, it is an *anisomyarian* clam; while if only the posterior adductor muscle remains, it is a *monomyarian* species. Reduction and loss of the anterior adductor muscle have occurred several times in clams that are attached by a byssus (fibers from the foot to a rock or other object). According to C. M. Yonge the advantage of this change lies in permitting the water current for feeding and respiration to enter the shell further above the surface to which the clam is attached.

More importance has been given to the structure of the hinge teeth. Even today some paleontologists consider that they are the single most valuable structure in determining the relationships of bivalves. In many groups the teeth do indicate degrees of relationship between species, but in the fresh-water Unionacea, for example, tooth types effectively encompass the total range of variation found in all other bivalves. The same tooth type has evolved in more than one lineage (convergent evolution). Despite this, the terms themselves are highly useful in describing basic variations. *Taxodont* (Fig. 4*a*) dentition has a large number of small alternating teeth and sockets extending along the dorsal hinge line. *Actinodont* (Fig. 4*b*) dentition has elongate teeth radiating from the umbos. *Heterodont* (Fig. 4*e*) dentition has the teeth differentiated into laterals and cardinals. *Isodont* (Fig. 4*c*, *d*) dentition is as in *Spondylus* (see also Fig. 3) or the less developed hinge of *Pecten*. Edentulous (Fig. 4*f*) taxa have no teeth at all. Numerous other tooth types have been named, and several other systems of naming and com-

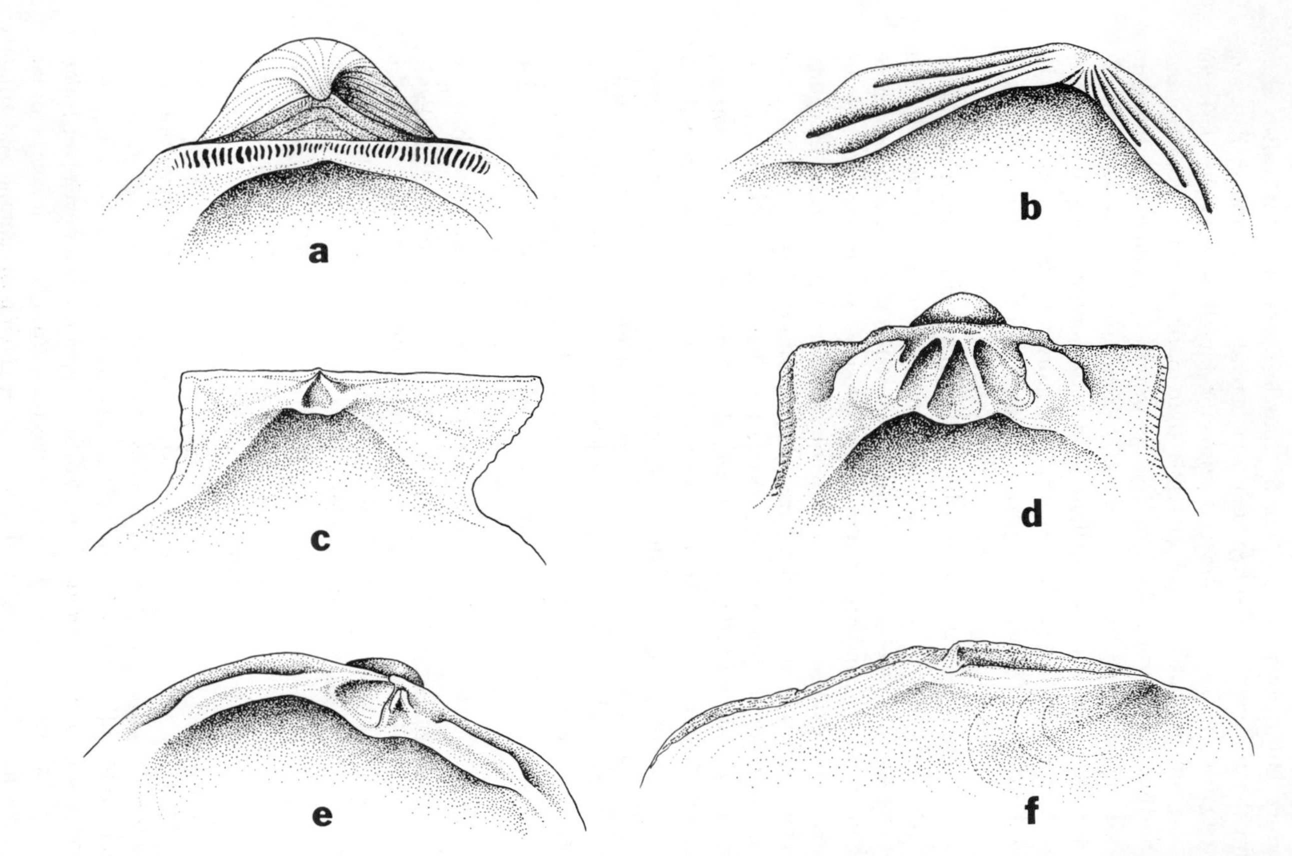

FIGURE 4. Hinge structures in (*a*) *Arca* (taxodont); (*b*) *Actinodonta* (actinodont); (*c*) *Pecten* and (*d*) *Spondylus* (isodont variations); (*e*) *Spisula* (heterodont); and (*f*) *Anodonta* (edentulous).

paring bivalve teeth are in use. Examples of some tooth types are seen in Fig. 4. If kept in the perspective that the teeth represent interlocking devices, subject to parallel evolution in different groups of clams, then the study of tooth similarities can provide much useful information.

Although frequently the shell will be regular in shape, with anterior and posterior portions in moderate balance on either side of the umbos, extreme elongation in such groups as *Barnea* (Fig. 1) and *Cultellus* (Fig. 5), truncation of shape in *Mya* (Fig. 5), or lateral compression as in *Isognomon* (Fig. 1) are not uncommon. Chosen for illustration because of its strong color and oddly deformed shape, *Corculum cardissa* (Plate 2, upper right) is proportionately as much wider than average as *Isognomon* is thinner. The monocolored, indented posterior end contrasts with the colorful, bulbous anterior end.

For many years there was a strict dichotomy in how bivalves were studied and in developing their classification. One school of scientists depended on the study of shell features, particularly in regard to the hinge teeth and muscle scars. The American paleontologist William Healy Dall epitomized this appgoach, while the Belgian zoologist Paul Pelseneer represented the opposite viewpoint, focusing on the structure of the bivalve gills and other anatomical features as the key to understanding their relationships. Within the past 30 years there has been a healthy tendency to use evidence from a variety of organ systems. But no two malacologists agree on how many orders and suborders of bivalves should be recognized, exactly how they are related, and what should be the limits assigned to each.

How the gills function in sorting food, where and how it is transferred to the stomach, how nonfood particles are ejected, and what is the pattern of feeding activity are among the features of clams studied most intensively. Even many textbooks cover these aspects of clam structure and function. Since clams are "headless" mollusks with their gills converted into a food sorting device, and their mantle cavity is enlarged to hold the entire body, the emphasis on studying these features is quite appropriate.

But here it seems more useful to focus on the patterns of clam diversity in relationship to habitat and structure. A clam's body is laterally compressed, with the mantle extending along each side either to or beyond the shell margins. A muscular foot of varying size can be protruded or withdrawn at the anterior end. Complicated sets of gills hang down on each side of the center line, filling most of the mantle cavity. Concentrated in the dorsal third of the shell is the visceral hump, which contains intestine, digestive gland, heart, kidney, and reproductive organs. Anus, excretory, and genital systems open into the mantle cavity. The head at the anterior end is reduced essentially to a mouth with two "labial palps" that help push

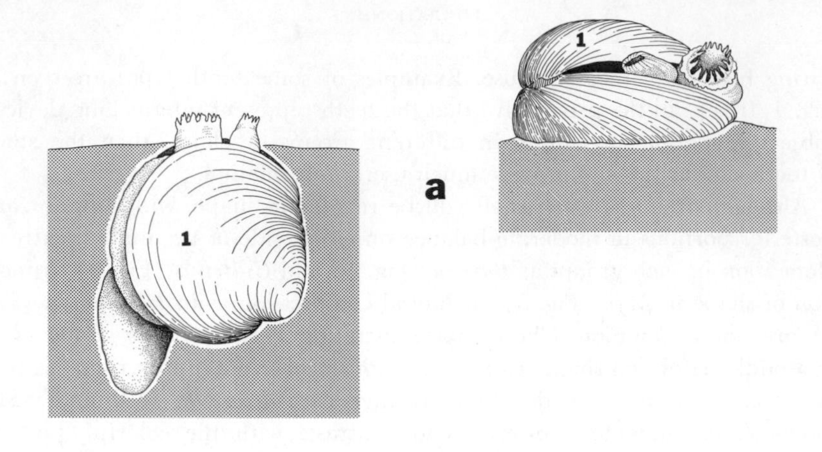

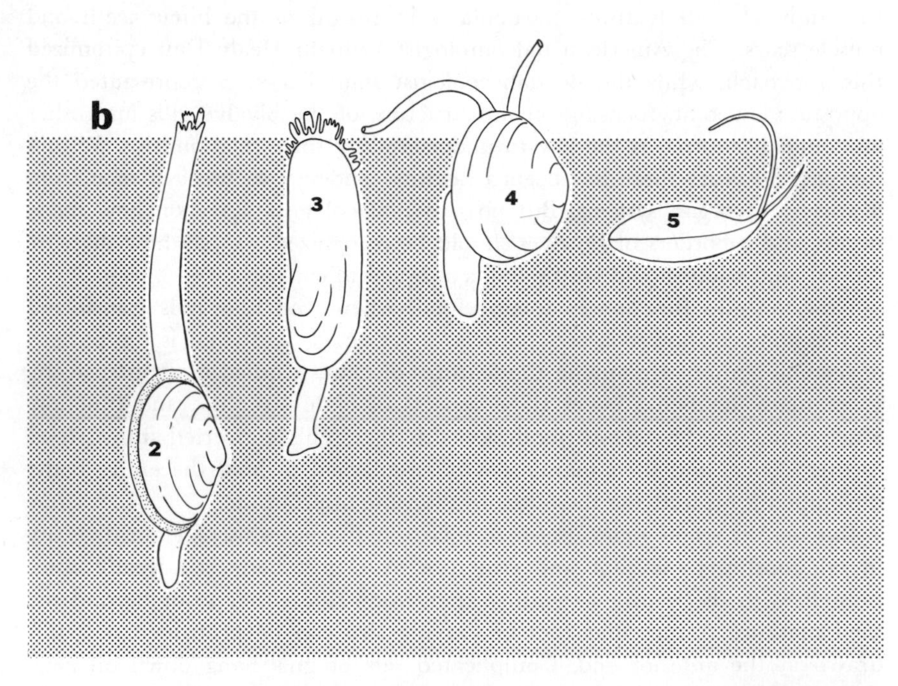

FIGURE 5. Siphon length and position in suspension and deposit feeding infaunal clams: (*a*) Dorsal (right) and side (left) views of surface dwelling suspension feeder showing the fringed inhalent siphon and smoother exhalent siphon at right; (*b*) siphon length and depth of burrowing in suspension feeders (1, *Periglypta;* 2, *Mya;* 3, *Cultellus*) and deposit feeders (4, *Scrobicularia;* 5, *Macoma*). After Meyer and Möbius (2–4) or original from color slides.

food into the digestive tract. Both buccal mass and radula are absent. Sensory equipment has transferred to the periphery of the body (siphons, mantle edges, and foot), where there is contact with the outer world. The mantle normally is fringed with tactile or chemoreceptors. In swimming forms such as *Chlamys* (Plate 3, upper row) the mantle edge has numerous tiny eye spots.

The gills serve as food sorting organs, except in the primitive nuculanid clams, where the labial palps can be protruded from the shell margin and are used to collect food, and the septibranch bivalves *Poromya, Cetoconcha,* and *Cuspidaria* that feed on the bodies of dead animals. A continuous current of water is involved. Water bearing oxygen and food enters the body through the inhalent siphon, usually at the posterior end, while water of lowered oxygen and food content, which may also pick up from the clam body wastes, undigested food, and sometimes eggs or sperm, exits at the posterior end of the body. The path followed by the inflowing water and extracted food particles is highly complex and differs greatly according to gill structure and habitat. In many clams the ventral margins of the mantle are partly fused, with an opening at the anterior where the muscular *foot* can be extruded and retracted, while at the posterior there are two tubelike organs, often formed by fused mantle tissue, the *inhalent* and *exhalent* siphons. The exhalent siphon is normally a simple tube, but often the inhalent siphon (Fig. 5*a,* right) has fingerlike lobes that filter out larger particles or at least are sensory and inform the clam to "stop pumping" until the potentially clogging particles drift away. In side view (Fig. 5*a,* left) a generalized clam is shown in normal feeding position, only partly buried in the silt.

The amount of water filtered by a clam varies with its feeding type and position in relation to the tides. Deposit feeders and those living below the tidal zone filter far less rapidly than those species that are exposed during low tides. *Crassostrea virginica,* the American Oyster, filters 4 to 15 liters of water each hour (a liter is slightly more than a quart) and the upper tidal zone *Mytilus edulis* (Chapter II, Fig. 6) up to 1.4 liters/hour. Both are exposed during some low tides. They must concentrate their feeding in the time when they are covered with water. In contrast, the large freshwater unionid clams, which are continuously covered with water, filter at about the same rate as *Mytilus,* although they have perhaps 10 times the body volume to nourish. Unionids can filter continuously, but *Mytilus* only a few hours a day. The sometimes intertidal but generally deeper water *Mya arenaria* (Fig. 5b, 2) has been measured as filtering at 0.96 liters/hour, while the deposit feeder *Scrobicularia* (Fig. 5*b,* 4) filters only about 0.3 liters/hour. Although these may seem to be relatively modest quantities, the cumulative effect of millions of individual clams filtering away makes the bivalves primary water cleaning organisms.

At this point an ecological distinction must be made. The description given above applies to a clam that digs into the sea bottom. Organisms that do this are members of the *infauna.* Those that move freely on the surface or attach themselves permanently to some object are referred to as *epifauna.* These modes of life involve major changes in the clam's structure. To some extent these adaptations are related to taxonomic groupings, but by no means exclusively.

Infaunal species include shallow and deep burrowers, deposit and suspension feeders. Shallow burrowing suspension feeders usually have a shell with strong hinge teeth, and often well-developed sculpture, a large foot, and short siphons as in *Periglypta* (Fig. 5*a,* 1). The animal sits just below the surface, so that in shallower water, the normal small wave movements partly uncover or cover the clam. In somewhat deeper waters are found suspension feeders such as *Mya arenaria* (Fig. 5*b,* 2). Their shell and foot remain a considerable distance below the surface. The siphons are long and partly to completely fused. If disturbed, the clam can withdraw the siphons below the surface, but only a small part of the distance to the shell. The siphons cannot be pulled back into the shell, which cannot be completely closed and has relatively small hinge teeth. Such deep burrowers can reach very large size. The West American geoduck (pronounced "gooeyduck"), *Panope generosa,* may weigh 12 pounds, have over 3-foot siphons, and an 11 inch shell weighing almost 2 pounds. It is not dug up by collectors, it is *excavated.*

The European *Scrobicularia plana* (Fig. 5*b,* 4) is an infaunal deposit feeder. It lies near the surface, often with part of the shell exposed. The siphons are long and clearly separated from each other for their entire length. The inhalent siphon sweeps over the surface of the bottom, pulling in particles from the layer of deposited organic matter. Somewhat in contrast, the West American *Macoma nasuta* (Fig. 5*b,* 5) is moderately deeply buried on its side, with the inhalent siphon protruding to sweep up deposits and the exhalent siphon either poked to the surface or partly retracted. Many species that live beneath the surface will build temporary tubes or holes up to the surface and then retract the siphons. Water currents flow through the open holes. Infaunal deposit feeders tend to move about more since they can exhaust local food supplies, whereas the suspension feeders have food circulating in the water itself. They are much more sedentary.

The process of burrowing involves foot, shell, and siphons. As outlined by Trueman, there is a series of actions occurring in sequence. The foot first extends downward in a probing motion and then expands into an anchor; then the siphons close to prevent any water being ejected; third, the valves are closed rapidly by the adductor muscles effectively expelling water from the ventral margin; this is immediately followed by contraction of foot retractor muscles, which pulls the clam shell and body downward toward the anchored foot; finally, the adductor muscles relax and the ligament will gape the valves apart. The water expelled ventrally in the third step creates momentary turbulence that loosens the sand or mud particles just beneath the shell. At the same time, the retracting foot stalk muscles pull the shell downward, displacing the loose particles up alongside the valves into the space that was just vacated by the closing valves. As the valves gape open, this displaces the material laterally and up over the top of the shell. This sequence of events probes, excavates, displaces, and then buries the clam in small to large increments. Rocking motions of the shell back and forth also aid in vertical descent. The prominent shell sculpture found on the shells of many infaunal clams specifically aids the burrowing process through its interactions with the bottom sediments. The many ridges, bumps, or scales seen on clam shells are not merely decorative. In a few cases, the shell of burrowing species will have long spines, as in the West Indian *Pitar dione* (Plate 2, upper left). These have been interpreted as providing protection from predators for the siphons. The spines extend on either side of the siphons and would make it difficult for a fish to nibble. They also could serve to stabilize the shell in a shifting bottom.

Some species, such as *Cultellus pellucidus* (Fig. 5*b*, 3) are extremely quick burrowers, the slender shell and elongated foot being adaptations for rapid disappearance. Such "razor clams" have evolved in several families. Species of *Solen* and *Ensis,* for example, can burrow almost faster than a collector can dig.

Life for epifaunal species presents a different set of problems. As permanently attached species, they must be suspension feeders, but if they live in the tidal zone, then feeding is possible only when they are covered by water. Attachment can be by cementing part of the shell to a rock, piece of wood, or other fixed object, by boring a hole in rock or wood that the animal then occupies permanently, or by secreting thin fibers (*byssal threads*) from the foot to a solid object. Erle Kauffman has recognized several types of epifaunal bivalves, each with a distinctive shell form: byssate free swinging species such as *Isognomon* (Fig. 1); byssate closely attached exposed forms such as *Mytilus edulis* (Fig. 6 in Chapter II); byssate nestlers such as *Arca* (Fig. 4*a*); byssate fissure dwellers such as some *Chlamys* (Plate 3, upper right); cemented forms such as oysters (Fig. 7 in Chapter IV) and *Spondylus* (Plate 2, lower right); and free swimmers such as most *Chlamys* (Plate 3, upper right). Borers generally are classed as infaunal mollusks, but the difference between a shell cemented to the outside of a rock and a clam making a hole in the same rock seems less than between the rock borer and a mud burrower.

The thin streamlined shells of *Isognomon,* the bulbous smooth *Mytilus,* abruptly angled *Arca,* and heavy shells of *Spondylus* all reflect different current conditions in their environment. The cluster of narrow *Isognomon* lives attached to projecting objects in shallow subtidal waters. The swirls of current continuously move them back and forth, their narrow width serving as an air foil in the currents. In contrast the tidal zone *Mytilus* can occur in dense colonies on rocks subject to the pounding of waves. Their more obese form plus very short and strong byssal attachment help minimize side to side movement by water currents. *Arca* lives in crevices of rocks and coral. Its broad top (Fig. 4*a*) is well adapted to blunting the pressure of a wave. At first glance, the heavily spinose shell of *Spondylus* would seem ill adapted to the rock surfaces on which it is cemented, but these spiny protrusions give coral nibbling fish an untasty mouthful and also when touched serve to alert the clam to the approach of potential danger. One valve of *Spondylus* is attached, while the upper, usually more heavily spined valve, is free to be moved. It must, however, be held up against gravity in order to stay open. At the same time it is subject to lateral pressures by shifting water currents. The far stronger hinge teeth of *Spondylus* (Fig. 4*d*) compared with the related *Chlamys* and *Pecten* (Fig. 4*c*) reflect this greater stress. In addition, the center part of the ligament, termed a *resilium* because it is internal, must be very large and powerful in order to gape the valves open.

The byssus is secreted by glands in or near the foot and usually consists of a multitude of fibers binding together the foot and the object to which it is attached. The capacity to form this structure is found in many families. Sometimes (in *Mya,* for example) a byssus is used by a young larval clam to secure itself temporarily to small pebbles, even though later it becomes a sediment burrower. For those species that live permanently attached, a plentiful supply of food is essential. The rich shallow continental shelf areas or the island coral reefs of the tropics are the primary regions where attached clams are found in great variety. Many live in or just below the tidal zone. In addition to battering by waves and wave-tossed rocks these clams run the danger of drying out if exposed to the air. Being permanently fastened, they cannot move about. Siphons are short, rarely extending any further from the shell than do those of *Mytilus* (Fig. 6, Chapter II), and the foot is usually atrophied. The shell is often thick and solid, and frequently distorted in shape by local factors.

Boring species use a variety of means to make a home for themselves. Such clams as *Lithophaga* secrete a weak acid from their mantle, then mechanically erode the weakened rock by rotating their shell. Naval or ship worms, members of the Family Teredinidae, bore into wood by using their shell as a file. Pholadid clams use mechanical abrasion alone to bore into soft rocks.

Attached and boring clams differ only in degree. They are stationary for most of their life, where most other clams move about occasionally. For all clams, initial major dispersal comes when they are larvae. Marine clams reproduce by releasing sex products from the genital pores into the mantle cavity. Fertilization either takes place there or the male and female gametes meet in open water after being pumped out of the exhalent siphon. Apparently there are very subtle "environmental triggers" or chemicals given off by the clams, since members of the same population often will "spawn" all at once. If spawn is introduced into a colony, then members of that group will begin spawning. The fertilized young goes through a planktonic larval stage, during which it can be transported many miles by currents. After a variable period of time, the larva settles toward the bottom. If it reaches a suitable type of bottom immediately, it will lose larval characters and transform into a miniature of the adult. It can, however, delay this transformation for a while if the bottom type is wrong. But this delay period is limited. If it does not soon find an appropriate soft bottom or place to attach, then the larva will transform and die on a bottom where it cannot live. Clams produce millions of young. Most are eaten in the plankton, many do not find an appropriate bottom type, and many others will be eaten before becoming adult, become stranded on the shore by a storm, or smothered by a layer of silt. Only a very few survive to reproduce.

Larval clams have only limited self-movement, depending on cilia to shift their positions, but a few adult clams can swim rather rapidly and effectively by opening and shutting their shells. The file shells (*Lima* and scallops (Family Pectinidae) both swim, but the scallops are particularly effective. A single huge adductor muscle ("scallops" in a restaurant are this muscle) opposes a shell valve gape of up to 30°. They can move forward, upward, or make a special escape movement by shutting and opening the valves. Two Australian *Chlamys* (Plate 3, upper left) have their valves gaped open about halfway. In keeping with their free moving life, their sensory equipment has become elaborated. Not only are sensory papillae on the outer mantle edges, but a series of bright blue spots (Plate 3, upper right) represent color from the light reflecting portion of eyes that are capable of detecting movement, but not shapes. The fringes on the inner edges of the mantle serve to prevent sediment from entering the mantle cavity and fouling the gills.

Until recently it was believed that a genus of clams found off Western

North America, *Chlamydoconcha,* was a permanently swimming member of the plankton, with completely internal shell. A study issued early in 1973 concluded that this genus was based on exceptionally long-lived larvae. It is not yet known to which adult clam these larvae belong, but the absence of any reproductively mature examples of *Chlamydoconcha* strongly suggests that this conclusion is correct.

Other remarkably varied adaptations are shown by the "watering-pot shells" of the Family Clavagellidae, the limpetlike species belonging to erycinid genus *Ephippodonta,* and a peculiar mytilid, *Fungiacava eilatensis,* that inhabits the body of certain corals. They also indicate the complexity of relationships between different organisms. Even more specialized habits have been recorded for other little known species of leptonacean clams. More than 30 genera of the Superfamily Leptonacea live in close association with other animals in modes of life ranging from symbiotic to parasitic. Only brief mention can be made here to details of these aberrant clams.

The most common watering-pots, such as *Brechites,* are burrowers in soft mud or sand, while other clavagellids, such as *Clavagella,* attach themselves to small rocks. In both genera the true shell is a remnant on the sides of a 3 to 9 inch long tube. The tube is up to 2 inches in diameter and very thin. The animal filter feeds through siphons extending to the open end of the shell which protrudes slightly above the bottom surface. The animal's biological anterior end is placed downward in the free-living *Brechites,* and is capped with a peculiar perforated disk that looks exactly like the expanded head of a gardener's watering-pot, hence the common name of watering-pot shell. Water movements through the disk may help the animal embed itself further into the sea bottom and could serve as a means of pulling potential food particles into the body area.

Species of *Ephippodonta,* which live in ocean waters off both Australia and Japan, are $\frac{1}{4}$ to $\frac{1}{2}$ inch clams that seem to mimic limpets in their activities and mode of life. Later (p. 115) I discuss the "bivalved gastropod," a tiny snail that has a very clamlike shell (Chapter VII, Fig. 5), so it is appropriate to mention "snaillike clams" here. The two valves of *Ephippodonta* are normally held open almost 180° to form an umbrellalike cover for the body. If turned over, the shell gape can be narrowed to about a 70° angle as a means of aiding the animal in righting itself. Normally the shell is held wide open. The animal creeps with an "inch-worm" type of locomotion through alternate contractions and expansions of the foot, whose anterior tip is modified into a disk that serves as a temporary holdfast organ during locomotion. The mantle is expanded in the Japanese species, but in Australian taxa it is reflected to cover at least part of the shell externally. The inhalent siphon protrudes in front of the foot anteriorly, while the exhalent siphon opens at the posterior end of the animal. Most finds of *Ephippodonta* have been in association with sponges or shrimp burrows. This probably is not a coincidence, but whether this indicates a symbiotic relationship where both organisms benefit (mutualism) or a case where only one benefits (commensalism) is unknown.

The kind of relationship is much clearer in the case of the mytilid *Fungiacava eilatensis*. Described from the Red Sea in 1969 and also reported from the Marshall Islands, this species lives inside the skeletons of fungid corals. The clam excavates a cavity in the coral's skeleton by means of chemicals secreted by the mantle tissue. It lies ventral side up in this cavity, with its siphon extending through the body of the coral itself into the coral's large central digestive cavity. Siphons of two or more clams may share the same opening into the cavity. *Fungiacava* does not compete with the coral for food. The coral feeds on large animals, while the clam's stomach contains quantities of microscopic protists called zooxanthellae that are mutualists of the coral. The clam is thus a commensal of the coral which preys on a mutualist of the coral. The shell of *Fungicava* does not serve a protective function and is modified in a number of ways. It is very thin, the hinge teeth are absent, the calcareous shell layers are greatly reduced in extent, and the mantle reflexes to cover it completely. Despite examining thousands of corals in museum collections, only rarely have specimens of *Fungiacava* been spotted, suggesting that this is indeed a rare and unusual clam.

Symbiosis is the technical term for organisms that are closely linked together. The relationship can be that of a parasite to its host, with the former draining nutrients from the latter; commensalism, where one organism benefits from the activities of the other; or mutualism, where both organisms benefit from the association. The classic example of the latter is a lichen, which is a combination of an alga and a fungus for the benefit of both. Deciding how to define the relationships of the unusual bivalves described above is difficult in the absence of more evidence, and serves mainly to illustrate the complexity of life.

No survey of bivalves can be concluded without mention of the largest species, the giant clams of the genus *Tridacna*. Found in and near coral reefs of the Indo-Pacific, they are characterized by an enormously thickened and colorful mantle. The shell has been twisted around so the umbos and hinge are underneath, with the shell gape on the upper side exposing the mantle. The mantle and siphons face up. The reason for this lies in the peculiar biology of *Tridacna*. It is a true "farmer," with colonies of algae living inside the mantle. Lenslike hyaline organs in the clam's mantle focus sunlight into the mantle layers, providing the sunlight needed for several layers of algae to grow. In addition to sifting food from the water, giant clams thus grow and harvest a portion of their own food. Their dependence on this farming is indicated by the fact that they are limited to waters shallow enough for sunlight to penetrate. The largest species, *Tridacna gigas,* is known to reach a shell length of 54 inches, and has a shell weight alone of 507 pounds. The weight of the animal from this monster is unknown. The smaller *Tridacna squamosa* (Fig. 6), characterized by its cuplike sculp-

FIGURE 6. A small (8 inches) specimen of the Philippine Island giant clam, *Tridacna squamosa.*

ture, lives in protected crevices on the surface of coral reefs. The only other clam known to raise algae is *Corculum cardissa* (Plate 2, upper right).

These then are the bivalves, a headless group of mollusks that filter food from water passed across their enormously enlarged gills. Within this limitation of feeding, they have diverged into deposit and suspension feeders; deep and shallow burrowers in the sea bottom; byssal attached species in a variety of habitats; forms cemented onto rocks, tree roots, or other shells; borers into rocks and wood; and even a few free swimming species. Because food suspended in the ocean water is constantly being circulated by the tides and currents, filter feeders could experiment with a number of abodes. Clams have done just that, devising different ways and picking different places to sit and wait while pumping for their dinner.

The massive treatise *Bivalvia* by Fritz Haas, the excellent review of clam biology and classification by Cox et al. in the *Treatise on Invertebrate Paleontology,* and the review of shell form and life habits by Steven Stanley all present far more detailed and comprehensive accounts of clam biology than this summary.

VI
Gliding Browsers

Wherever the land meets the sea, in salt marsh, rocky cliff, sand beach, or river delta, a rich profusion of microscopic life flourishes. Nutrients wash down from the land, or upwelling currents from the ocean deeps lift the chemicals of life from deposits accumulated over the eons. These combine to support areas of productivity equal to that of the finest agricultural lands. A profusion of filter feeders feast on the microorganisms, and in turn the filter feeders provide a rich harvest for mobile predators. Annelid worms, starfish, flatworms, fish, and snails hunt filter feeders and each other in constantly changing patterns of abundance and interactions.

While the clams remained filter feeders and explored places and ways of attachment and burrowing, other mollusks moved about the shore after food. The chitons, protected by their girdle and armored plates, stayed as slow browsers on rocky shores. The sensory equipment that extends up through their shell compensated for the problems of clamping to a rock. Snails retained a gliding foot, intensified their anterior sensory equipment, perfected the shell as a place of retreat when in danger, and experimented with food sources and methods of feeding. In time they became ecologically the most diverse class of animals. Snails are equally at home in the ocean, in bodies of freshwater, and on land. Although the perhaps 67,000 species of snails and slugs are dwarfed in numbers by the possibly 3,000,000 living species of insects, there are only a handful of marine insects. Crustacea are abundant in the sea and freshwaters, but only the common land isopods (pill or sow bugs) and a few amphipods now in tropical forests came onto land. Mammals are common on land and in the ocean, but the number of species, about 4000, is a fraction of that represented by molluscan or arthropod diversity.

Because the gastropods are so diverse in species and habitat, a larger portion of this book is devoted to them than to the other classes. Three great subclasses of gastropods are recognized, based on anatomy, reproduction, and habitat exploitation. The Prosobranchia is the most primitive of these, with separate sexes, and almost always a heavy shell. They are most abundant in the ocean, although they also colonized freshwater and land habitats. The Opisthobranchia was derived from the Prosobranchia, but the species are mostly hermaphroditic, generally have a reduced shell or none at all, and virtually all are marine dwellers. The Pulmonata also were derived from the Prosobranchia, are hermaphroditic, lung breathers, and live mostly on land or in freshwater. Marine snails and slugs are reviewed in this and the following chapter, land snails and slugs in Chapters X through XII, freshwater species of all groups in Chapter VIII, while Chapter IX focuses on the feeding apparatus of snails.

In "designing" a basic mollusk (Chapter III), the last two stages (Fig. 5*d,* e) are charging off more in the gastropod direction, reflecting perhaps my own partiality to snails, but also focusing on the "progressive" criteria of increased mobility and formidable anterior sensory area. With only a few modifications, Fig. 5*e* would make a fine ancestral snail. The head is equipped with several sensors, it is protruded beyond the shell for searching activities, there are distinct anterior and posterior shell cavities, the foot is slender and thus adapted for gliding, but there are too many gills and excretory pores. Most of all the animal is not "torted." As mentioned earlier, "torsion" occurs in an early larval stage of a snail. The visceral hump and mantle cavity rotate up to 180° counterclockwise, bringing the posterior mantle cavity and associated organs to lie anteriorly above and behind the head, pointing forward.

The origin and initial functional significance of torsion have been debated hotly for years, with opinion divided as to whether the larva or the adult was helped more. A. Lang in 1900 suggested that torsion in conjunction with spiral coiling would help balance a high shell. A. Naef in 1911 proposed that the early gastropod had a shell coiled over the head like that of *Nautilus* (Plate 1). A forerunner of such a coiling pattern was subsequently found in *Neopilina* (Fig. 4 in Chapter II). W. Garstang in 1920 theorized that torsion aided the larva by allowing the head and foot to be withdrawn into the mantle cavity first, while in the "pretorted" mollusk the tail was withdrawn first, leaving the head exposed to harm. John Morton in 1958 hypothesized that the anterior shift of the mantle cavity permitted the mantle area to receive clean water rather than silty water stirred up by the foot. Michael Ghiselin in 1966 suggested that torsion helped the larval snail settle to the bottom and balance its shell. T. E. Thompson in 1967 criticized Garstang's theory, while A. J. Underwood in 1972 suggested that the two 90° rotations in torsion might have evolved independently.

Torsion, like the wing of an insect, a bird, or a bat, obviously has advantages to its possessor, once fully developed. It is very difficult to suggest why torsion got started and how it helped the snail in its initial stages. A fully torted snail has better protection for the head of both adult and larva, the larva can settle more easily to the bottom, the snail can sample cleaner water that originates and enters the mantle cavity farther above the surface on which the snail is crawling, and its chemoreceptors can sample the environment ahead of it more effectively.

Before speculating on the origin of torsion, it is necessary to review briefly the origin and advantages of a spiral shell, and to examine structures found in the earliest fossil snails. The living monoplacophoran, *Neopilina,* and the extinct bivalve group Rostroconchia demonstrate that a spiral shell can evolve in the absence of torsion. I have hypothesized above (pp. 41 to 47) that the spiral shell of a primitive mollusk served to protect the mollusk's head by being "rocked" forward and over it. This function necessitated that there be "empty space" at both the anterior and posterior parts of the shell. Also there must be retractor muscles attached to the shell in such a way that they could pull the shell forward. These muscles had to extend from the shell forward to attach on the body wall near the head. Presumably simple inertia during crawling or the overbalance of having the heaviest part of the weakly spiral shell lying posterior of the snail's middle served to shift the shell into a "normal position" once the danger that led to the forward "rock" had passed.

The primitive mollusk probably had many pairs of gills and excretory openings as suggested earlier (Fig. 5*e* of Chapter III). A gradual reduction in the number of such pairs would make sense for an active hunter. Its own movement would increase the flow of water past the gills. Hunting animals generally have a streamlined shape compared with their plant eating relatives. Accelerated concentration of respiratory and excretory openings toward the posterior end of the body would have been an early facet of snail evolution. Eventually this reduction would reach a stage where there was a single pair of gills and a single pair of excretory pores clustered in the posterior shell cavity.

Such a weakly coiled shell and hypothetical "first snail" is shown in Fig. 1. It is untorted and deliberately drawn so that the shell resembles some

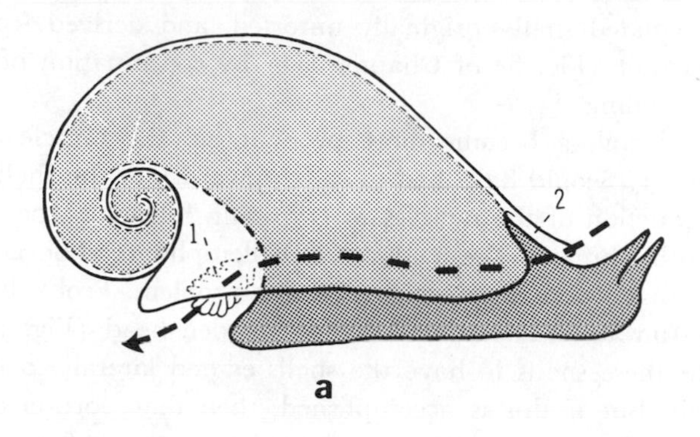

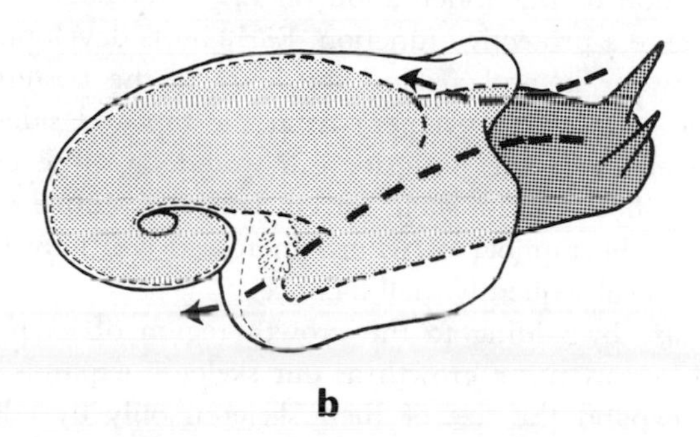

FIGURE 1. Hypothetical ancestral snail before torsion with coiled spiral shell, posterior pair of gills (1) and the future "anal notch" (2) serving to shield the head: (*a*) lateral view; (*b*) dorsolateral view. Arrows indicate direction of water currents used in respiration.

of the earliest known snail fossils, the bellerophonts. Preservation of these earliest snails is not very good. Classification of these fossils into the Order Bellerophontina may be lumping similarity of form arrived at independently through convergent evolution. At present we know of no characters by which they can be divided into several groups. The whole group is defined by having a simple planulately coiled shell. The bellerophonts lived from the Lower Cambrian to the Lower Triassic. There was only a single pair of muscle scars on the shell. While the first species had a simple shell edge, those from later geologic periods had an indentation or "anal slit" on the outer lip of the shell. These two features have been interpreted to mean that bellerophonts were "torted" snails. The apertual shape and rudimentary nature of the slit in earlier species suggest to me that bellerophonts might have been nonoperculated snails, originally untorted, and derived from the hypothetical ancestor (Fig. 5*e* of Chapter III) by continuation of the initial trend to spiral coiling.

As the spiral coiling became more pronounced, the muscle attachment for shell "rocking" would have had to shift upward into the shell and probably led to reduction of the muscles to two main retractors, paralleling the reduction in other organs. The tendency of bellerophonts to develop "flared" openings may have been connected with this problem. Probably the "anal slit" began as an accommodation for the extended head (Fig. 1). It is an advantage for these snails to have the shells extend laterally over the sides of the animal. But if this is accomplished, then that portion of the shell edge nearest the head must become indented so that the head can extend in front of the shell, yet preserve the lateral protection of the body by the shell. The formation of this indentation on the outer edge of the aperture initially would serve a protective function. Early in its development, it would act to channel water in and around the body to the posterior gills (Fig. 1b). This pattern of water movement from the top and sides of the head under the shell edges to the posterior gill would be quite effective. With only one or two whorls in the shell, an efficient and simple system existed. Problems arose as the number of whorls increased. Initially this would involve shell size and subsequently shell balance.

We grow in size by adding to the growth region of each of our many bones, followed by soft tissue growth as our skeleton expands in size. Snails and clams can expand the size of their skeleton only by adding material to the outer edges of the shell. While arthropods must shed their external skeleton at regular intervals, then build a new one to enable growth, snails and clams keep the same external skeleton, but must increase their size by expanding the edges. The parallel and downward growth of the clam shell makes this relatively simple. Snails, like scaphopods, have a problem in that growth requires elongating and expanding a cylinder. In the scaphopods, the slightly curved and gently tapering cylinder form causes no problems when burrowing in a mud bottom. For the snail, spiral coiling reduces shell length, but particularly with larger size, the basic spiral shape presents a height problem for the crawling snail.

This is partly a matter of balance. A planulately coiled shell of more than two or three whorls carried as in Fig. 1 would be easily tipped to one side or the other by minor water currents. It also would extend upward a fair distance in relation to the width of the body and shell.

This latter feature would cause another practical problem. A basic defense strategy for any hunted organism is to retreat into a crevice that is too narrow for its enemy to enter. The lower the shell height, the narrower a crevice that the snail can enter. Hence reducing the height of a shell could have a selective advantage for survival of the snail. A spirally coiled shell with the apex extended laterally has a much lower total height at the same shell volume than one that was planulately coiled. The snail could crawl in further and would be at an advantage. But merely extending the spire laterally would unbalance the animal to the right (Fig. 2*a*). Tilting and rotating the shell for balance in "carrying" corrects this problem (Fig. 2*b, c*). If the original shell protrusion was to the right, then for balance the shell would be rotated clockwise (directional arrow 4) and shifted slightly upward (directional arrow 5). Accomplishing this shifts the pattern of water currents significantly. The distance water has to travel to the *right* gill is less and in a more direct line than currents to the *left* gill. With water tending to flow more quickly to the right gill, any counterclockwise twist to the body that brought both gills nearer the faster current would have a selective advantage. The closer the gills rotated to the front, the greater the advantage.

Thus the mechanical necessity to balance an asymetrically coiled spiral shell by partial *clockwise* rotation and upward tilting would have created an imbalance in water currents that could favor a *counterclockwise* twist to the visceral hump. This would have had many and substantial effects on the basic anatomy and behavior of the snail. The asymmetry of the shell position initially would have altered the pattern of protective "rocking." Instead of paired muscle attachments, selection would have quickly favored fusion of the muscles to one point of attachment near the center axis of the shell. As the right gill and its space area round it began to shift even slightly toward the right side, then the partial combination of anterior and posterior "shell spaces" toward the front would provide an enlarged frontal area under the shell. In these circumstances, and with the tilted angle of the shell, the singly attached retractor muscle system could begin to pull the head *back* as well as move the shell *forward*. It would take only a slight shift of the right gill counterclockwise and the shell rotation clockwise for all the postulated advantages of torsion to begin operation. The multiplication of beneficial results as the process proceeded would have accelerated the selective value of these changes.

When rotation to the front was completed, the right gill would have moved over to the left side, and vice versa. Subsequent lengthening of the "anal slit" in the shell would partially solve a problem that still exists in primitive living gastropods. Water enters both sides of the mantle cavity to cross the gills, but then it must be evacuated with body wastes. To have

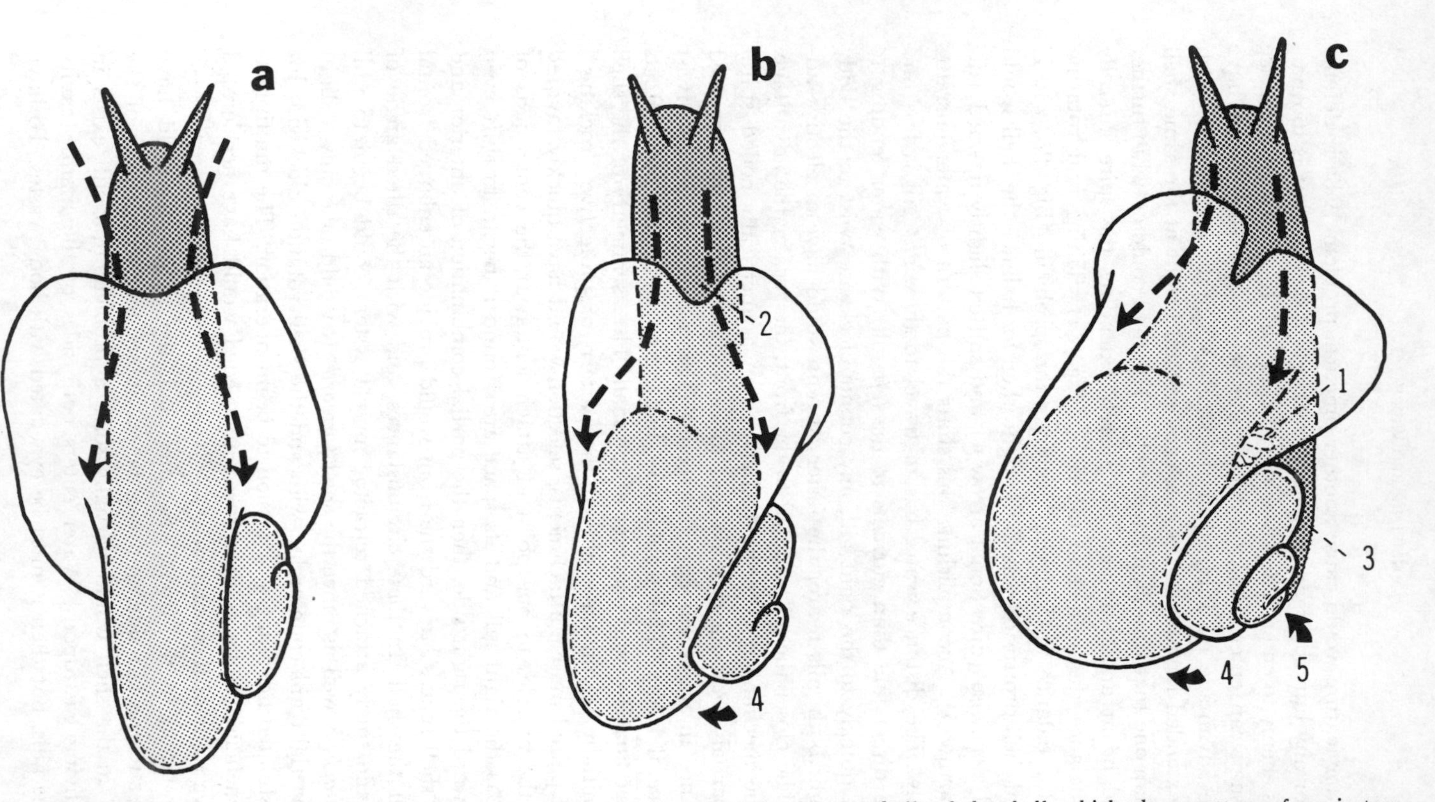

FIGURE 2. Stages in adjusting to spire protrusion by clockwise rotation and upward tilt of the shell, which alters pattern of respiratory water flow, favoring the right posterior gill. This would give strong advantage to any counterclockwise rotation of the gills toward the front of the body: (*a*) obvious imbalance in carrying shell caused by spire protrusion; (*b, c*) successive stages in spire protrusion and adjustments of shell position for balance. Structures are gill (1), "anal notch" (2), and foot (3). Clockwise rotation (4) and upward tilt (5) directions are shown.

it both enter and come out the front is rather inefficient. By deepening the anal slit, wastes and deoxygenated water can pass up and out through this "exhalent slit." Now the posterior mantle cavity lies in the "gastropod" position, facing forward, and combines with the anterior shell space to provide room so that the head and foot can be withdrawn completely. Complete retraction of the head and foot becomes possible by shifting the free muscles forward from just the midbody area to the head and anterior foot. Another result of the asymmetrical spiral shell coiling is often a more circular aperture. If the upper part of the tail is then strengthened above by conchin and eventually calcium, you have a "door" or *operculum,* capable of sealing the aperture behind the retreated animal. Such a predator can roam over the soft bottoms, retracting into safety if endangered. There will be paired organs (Fig. 6) in the pallial cavity, which now samples the incoming water for chemical signals, instead of relying only on the head. This then is an adequate portrait of a living archaeogastropod, such as the pleurotomarians and *Scissurella.*

I suspect that torsion started in a spirally coiled shell that had begun to coil asymmetrically in order to fit narrower crevices and became "tilted" to "balance" the shell while crawling. This favored water flow to the right gill and counterclockwise gill rotation. Once gill rotation had been completed, a "space retreat" anterior mantle cavity existed. Upon evolution of foot withdrawal capacity and formation of an operculum, we have a primitive archaeogastropod. Previously it has been hypothesized that torsion occurred before asymmetrical spiral coiling, but as outlined above the tilt correction needed to balance an asymmetrical coiled shell could have provided the initial impetus for what eventually was almost 180° torsion. No macromutation is required, only gradual evolutionary change. Foot and head withdrawal and an operculum came later. Shells alone do not enable us to say whether the bellerophonts with deep anal slits are directly related to the early ones, or whether they are secondarily simply coiled. Quite possibly the changes discussed above occurred more than once.

The most primitive existing gastropods are those with two sets of pallial organs—auricles, gills, kidney openings—plus a many-toothed radula (see Chapter IX). These belong to the Order Diotocardia or Archaeogastropoda depending on which name you prefer. Those with only one set of pallial organs are placed in the Order Monotocardia, which is younger in age and more advanced in structure. The small *Scissurella* and the large deep water pleurotomarians with their spiral shells and deep anal slits probably come closest of all archaeogastropods to resembling the basic gastropods. Fretter and Graham distinguish three major groups within this complex, two of which evolved into limpets or limpetlike shells. Embryological studies of these limpets show that they start out with a spiral shell and operculum, losing both during later development. The limpets are secondarily evolved from spiral shelled ancestors and are not primitively cap shaped.

Like chitons, limpets generally inhabit rocky shores or the fronds of seaweeds. Their conical shell, like that of a barnacle, is designed to withstand maximum wave action without the animal becoming dislodged. Instead of having a flexible girdle that enables the body to cling tightly to the rock surface, limpets, by a combination of weak erosion to both rock and shell edge, fit tightly against the rock. Frequently a limpet will have a "home base" to which it returns after each foray in search of food. They are browsers on encrusting organisms, spending much time on "home base," and usually foraging briefly. Prominent among these are the keyhole limpets, Family Fissurellidae, with either an anterior slit or an apical hole to aid in waste disposal (Fig. 3*a, b*). A fringe of tentacles from the sides of the body extend outward as an *epipodium,* providing tactile and chemoreceptors

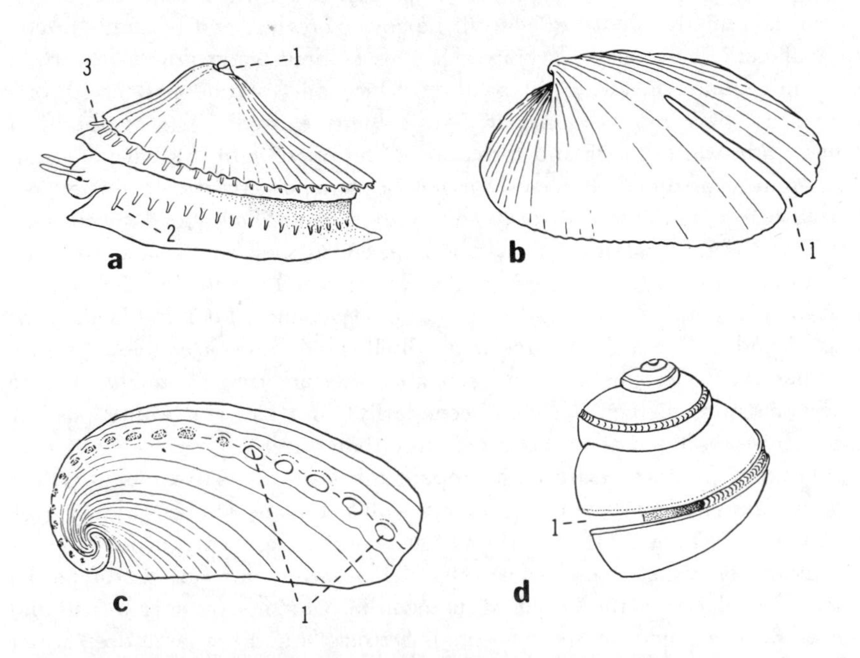

FIGURE 3. Anal slit and shell form variation in living archaeogastropod snails: (*a*) *Fissurella;* (*b*) *Emarginula;* (*c*) *Haliotis asinina;* (*d*) *Scissurella crispata.* Structures shown are anal slit (1), epipodium (2), and mantle papillae (3). After Forbes and Hanley (*a*) or original. The living animal of *Haliotis asinina* is shown in Plate 4, top.

on all body edges. In the abalones, Family Haliotidae, a spiral shell remnant is left, the anal slit appears as a series of holes in the shell (Fig. 3*c*), and the epipodium (Plate 4, top) has been elaborated into a series of ovoid dorsally protruding flaps. Abalones have a very broad and powerful foot for clinging to rocks. A few species reach 12 inches in length. Abalones are extensively used as food by man and, off Western North America, by sea otters also. The latter are skilled at flipping an abalone loose from a rock with their paws, while men frequently need a crowbar to perform the same trick.

The epipodial fringe of keyhole limpets and abalones is lost in true limpets of the Family Acmaeidae and Family Patellidae. Either one or both of the gills are lost. They substitute protrusions from the mantle edge as gills and others to serve as sensory tentacles extending past the shell edge. These limpets have the same conical shape as the keyhole limpets, but lack any slit or puncture in the shell and can live far higher in the tidal zone. When clamped down on their "home base," virtually no water will be lost. Many limpets in South Africa reach 3 or 4 inches in length and are prized for food, despite having a hard radular ribbon that is up to three times longer than the shell. If the animal is eaten by man, these radulae can catch on the soft projections of our intestine, gradually accumulating into a knot of radulae. In time this may produce "bleeding ulcers" requiring surgery. "True limpets" are among the hardiest inhabitants of the shore zone, but like the less modified keyhole limpets and abalones, they are evolutionary "dead ends." Successful on rocky shores and more diverse than chitons, there is no evidence that more advanced gastropods evolved from any of the limpet groups.

The remaining archaeogastropods, members of the superfamilies Neritacea and Trochacea, retain spiral shells. They show a tendency toward basic improvements of the mantle cavity which resulted eventually in major progressive changes in the gastropods. The Trochidae, which have a horny operculum, and the Turbinidae, which have a calcified operculum, are small to large marine species. Such shells as *Astraea* (Plate 3, lower left) and the spectacular *Guildfordia triumphans* (Fig. 4) are turbinids. Frequently the empty shells of these snails are used by hermit crabs, such as the Australian *Dardanus megistos,* shown in Plate 4, third from top, occupying a large *Turbo* shell. The neritaceans are smaller in size, and are tidal zone, subtidal, supratidal, freshwater (Neritidae) or terrestrial (Helicinidae) in habitat. Shells of neritids are also used by hermit crabs, species of *Coenobita* (Fig. 5) having the claws modified to fit the shell opening as tightly as the operculum of the original snail.

All higher archaeogastropods show varying modifications of the pallial

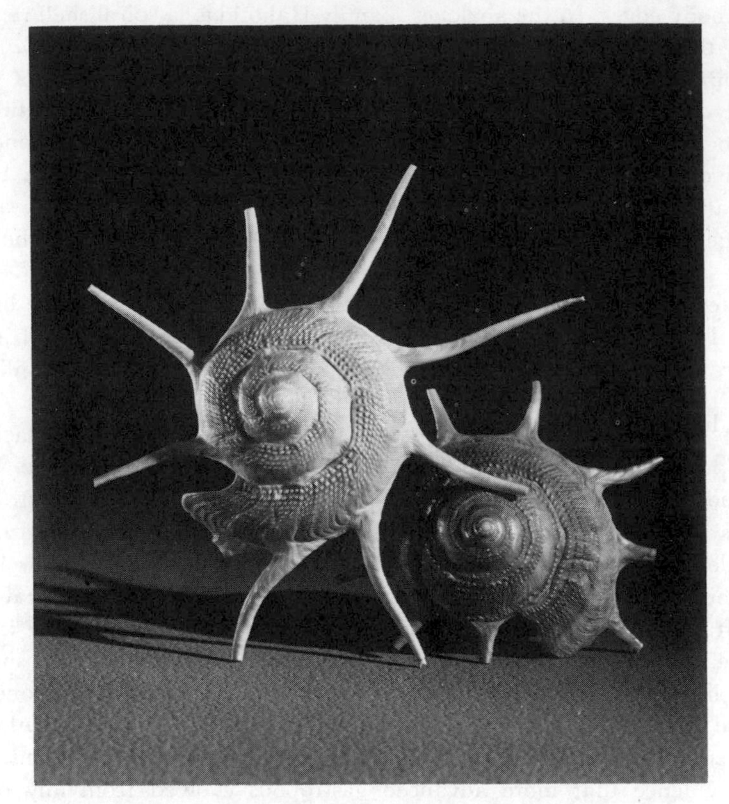

FIGURE 4. A deep water turbinid snail, *Guildfordia triumphans,* from Japan. Courtesy Field Museum of Natural History, Chicago.

region. In the primitive taxa with two gills, water enters both sides and then must be ejected, together with body waste products. While the dorsal anal slit does work, the small size of the slit in relation to the mantle cavity opening itself limits the rate of water flow. A continuous current with equal entrance and exit areas would be more efficient. In addition, the anal slit presents a vulnerable spot for attack by predators. All of the Trochacea and Neritacea show degrees of size reduction in the right gill, in addition to reduction or loss of the other right side organs. Because of shell tilting and rotation for balance, the left organs are nearer the entering water currents. In higher gastropods only the left side mantle organs remain (Fig. 6). Water enters on the left side, passes over the gills and a sensory organ called the *osphradium* (4), then goes to the back of the mantle cavity and

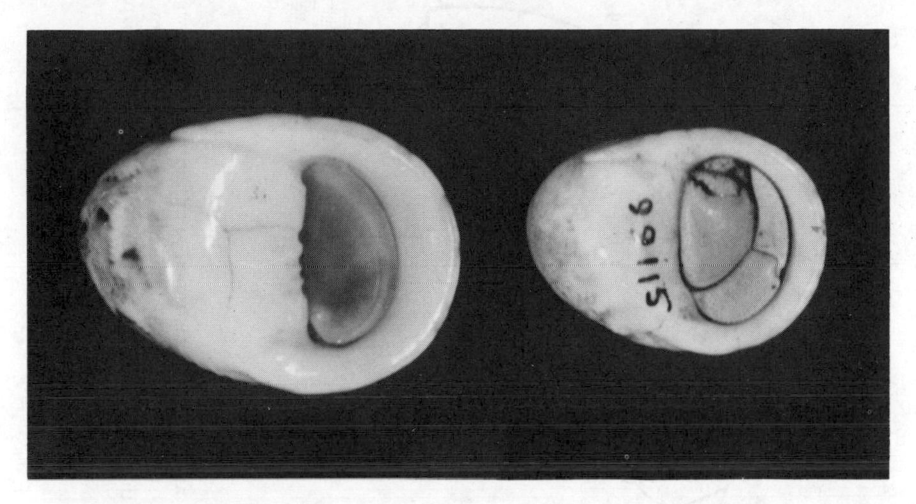

FIGURE 5. A hermit crab, *Coenobita* sp., withdrawn into the empty shell of *Nerita polita* (right). The crab's claws are modified in shape so that they plug the shell opening as tightly as did the snail's operculum (left).

passes the single excretory pore to pick up waste products, and finally turns forward along the right side. The anus and genital opening are shifted to the right front margin, so that just before the water current exits, it passes these pores, picking up and ejecting feces and sex products. There is thus a continuous and even flow of water coming in on the left side and going out on the right side. Reduction and then loss of the right mantle organs greatly increased the snail's ability to monitor what lay ahead and opened the way toward a variety of experiments in sophisticated predation.

At the same time, patterns of reproduction were changing. Fertilization in the higher archaeogastropods no longer takes place in the water, but inside the female. Eggs are laid singly or in strings, and much of the larval life passes within the egg capsule. In the more primitive members of the other prosobranch order, the "mesogastropod" complex, there are other advances. The many-toothed radula of the archaeogastropod is changed to one with only seven large teeth in a row. Except for eyestalks, the body lacks tentacular projections. Following Fretter and Graham, the higher prosobranchs, Order Monotocardia, are split into two suborders, the Taenioglossa or Mesogastropoda, and Stenoglossa or Neogastropoda.

The "mesogastropods" are, according to Fretter and Graham, in ". . . a general transition from a vegetarian and largely microphagous type of animal to one which has become carnivorous and to some extent parasitic."

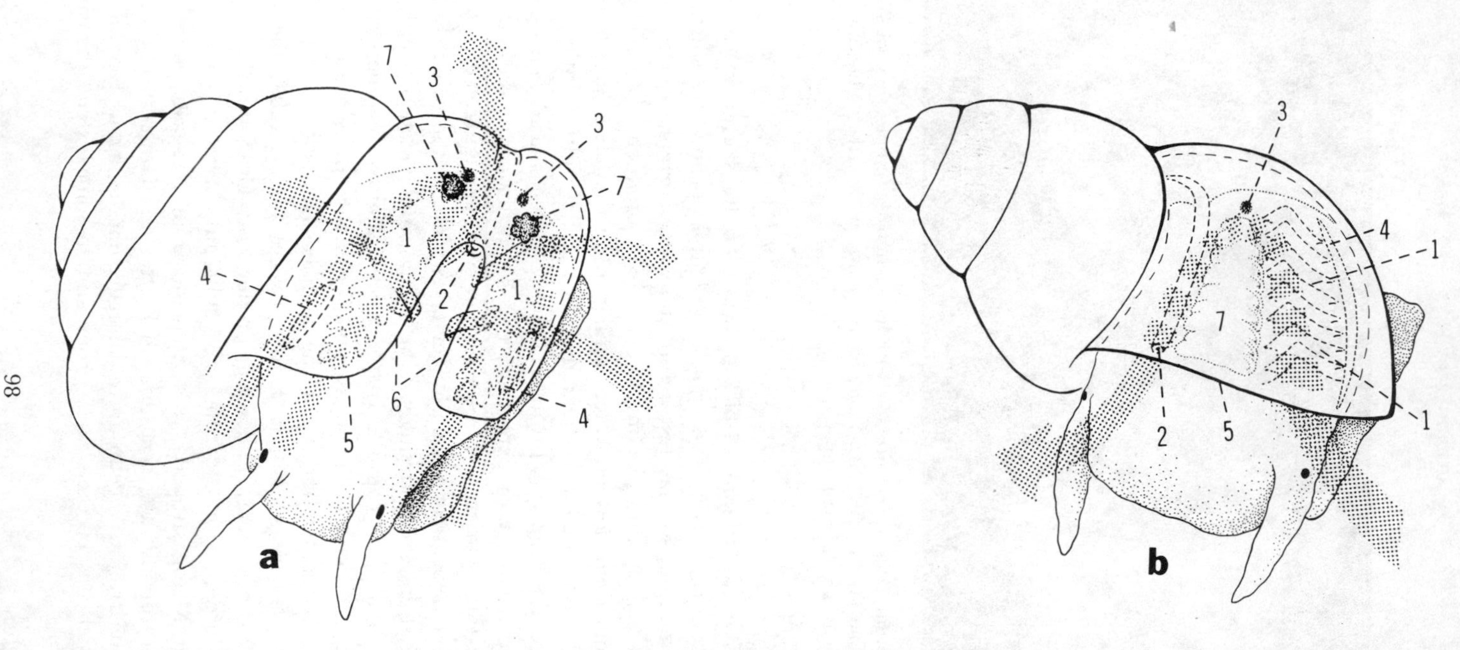

FIGURE 6. Structure and water current patterns in (*a*) idealized archaeogastropod and (*b*) mesogastropod snails. Structures shown are gills (1), anus (2), excretory pore (3), chemoreceptor (4), shell edge (5), anal slit (6), and hypobranchial glands (7). Arrows indicate path of water currents. Dotted structures lies against the inner surface of the shell and are viewed as if the shell were partly transparent.

Included are the majority of freshwater snails, many land species, such common seashore dwellers as periwinkles (Family Littorinidae), conchs (Family Strombidae), cowries (Family Cypraeidae), and most of the small to medium sized, elongated shore snails. A few groups of mesogastropods have become pelagic. *Janthina* builds a "float" of air bubbles which keeps it at the surface. It feeds on floating coelenterates, secreting a purple substance that may paralyze the victim, or may be used as a defense mechanism by the snail. Empty shells of *Janthina* are often found on beaches after storms. Other pelagic mesogastropods are the heteropods, swimming predators on animals up to the size of and including small fish. Moon shells, Family Naticidae, are sand dwelling carnivores that bore holes in clams, while the tun shells, Family Tonnidae, bore holes in echinoderms to feed on the animal. In contrast, groups such as the Turritellidae are burrowing filter feeders, and slipper limpets, Family Calyptraeidae, are attached or semiattached filter feeders on rocks or other shells. Worm shells (Fig. 7) are elongated, decoiled gastropods of shallow waters. A few families have become parasitic, eulimids as ectoparasites and entoconchids as internal parasites of echinoderms. Mesogastropods are a highly varied group. It is possible here to discuss only a limited sample of this diversity.

Perhaps the most familiar of the mesogastropods are the periwinkles, *Littorina,* of rocky shores. One scientist estimated that a square mile of rocky shoreline had 860,000,000 *Littorina.* In one year these animals would scrape up 2200 tons of material, only 57.2 tons of which would be organic matter. The rest would be indigestible skeletal material or bits of rocks. In order to accomplish this, *Littorina* has radular teeth heavily strengthened with mineral salts, and the radula itself is extremely long and fast growing, since teeth scraping rocks will become worn and chipped very quickly! A Bermudan *Littorina ziczac* with a shell slightly more than $\frac{3}{4}$ inch long had a radula over $6\frac{1}{2}$ inches long.

Littorinids also demonstrate clear niche specialization and variations in reproductive patterns. Several species in England live on the same shores, but are found on different parts of the shore. Their reproduction correlates with their habitat. *Littorina saxatilis* ranges from the midtide mark to several feet above the high tide mark. Its eggs are kept inside the female's body until they hatch, ready to crawl away and start scraping rock surfaces. *Littorina obtusata* is found on large algae such as *Fucus,* which live on the lower shore and are rarely exposed even at low tide. *L. obtusata* lays a gelatinous mass of eggs on the algae. The young pass through the normal swimming larval stage inside the egg, hatching as crawling young two or three weeks after the eggs are deposited. *Littorina littorea* lives in the lower half of the tidal zone, generally on rocks that are relatively free of encrusta-

FIGURE 7. Cluster of worm shells from Florida, *Vermicularia* sp. Courtesy Field Museum of Natural History, Chicago.

tions. The female lays pelagic egg capsules that hatch into a swimming larva six days later. *Littorina neritoides* sits in exposed rock crevices that are above the normal high tide mark, but which are inundated during very high tides or storms. It releases pelagic egg capsules on such occasions, but can last for at least five months without being doused by seawater. Littorinids in other areas of the world show similar zonation and variation in breeding types, but the British species have been studied more extensively.

In tropical and subtropical waters, conchs and spider shells (Family Strombidae) are common denizens of sand or mud areas with standing beds of algae. They feed on the algae fronds, and are in turn sought by man for use in conch chowder, and by various carnivorous mollusks. Spider shells (*Lambis*) have adult shells with spiny extensions of the lip (Plate 3, lower left) whose number and length vary with the species and age of the animal. All agree in having a shell that is very large and quite heavy. Presumably the extra heavy shell provides a certain amount of protection from other mollusks, although its tanklike trail through an algal patch is easy to follow. But this shell does cause the animal troubles in moving. The foot of *Lambis* (Plate 3, center left) is divided into an anterior spade-shaped *propodium* separated by a narrowed region from the posterior *metapodium* which terminates in a clawlike, highly modified operculum. The metapodium is on the far left of Plate 3, center left; the spade-shaped propodium can be seen just to the left of the right spine; while nearer the right shell margin the tubular *proboscis* and bright eyes are peeking out from the aperture. In moving (Fig. 8), the propodium (F) is stretched forward to just behind the proboscis (P) and pressed against the bottom. The metapodium is then pulled forward under the shell, and the operculum (O) pushed into the bottom. A sudden upward and forward heave by the snail lifts the heavy shell off the bottom and throws it forward about half its length. The proboscis (P) and propodium (F) are pulled forward, and then the process is repeated. In this way the snail proceeds by a series of lurching jumps. Well-developed eyes are borne on the tips of stalks branching from the long tentacles. The eye itself is brightly colored and has up to five concentric color rings developed. Usually the outer surface of a strombid shell will be a garden of algae and bryozoans or small sponges, so that the bright colors of the shell aperture and the animal itself come as rather a surprise.

In contrast the cowries (Family Cypraeidae) and their relatives are famous for their brilliantly colored shells. They are among the favorite groups of shells for collectors. What has been little appreciated is that the animal often is as attractive as the shell, but colored completely differently (Plates 6 and 7). These animals start life with an obviously coiled shell, but in the process of becoming adult, all but a tiny tip of the original spiral coiling is covered (Fig. 9), the aperture becomes quite narrow, and is eventually lined on both edges with toothlike projections. Cowries are generally predators on ascidians and will scavenge on available dead animals. In any one area a number of species may be found, but other than notes on habitat differences, there are virtually no data available as to how they specialize in terms of feeding or behavior.

Although it is an "egg cowry" (Family Amphiperatidae) and not a true cowry (Family Cypraeidae), *Calpurnus verrucosus* (Plate 6, upper right) demonstrates a clear picture of basic body structure. The broad foot can be extended laterally and well behind the shell margins. It is identifiable by the more widely scattered black spots. The mantle is divided into two flaps of tissue that can be extended up the sides of the shell to meet along

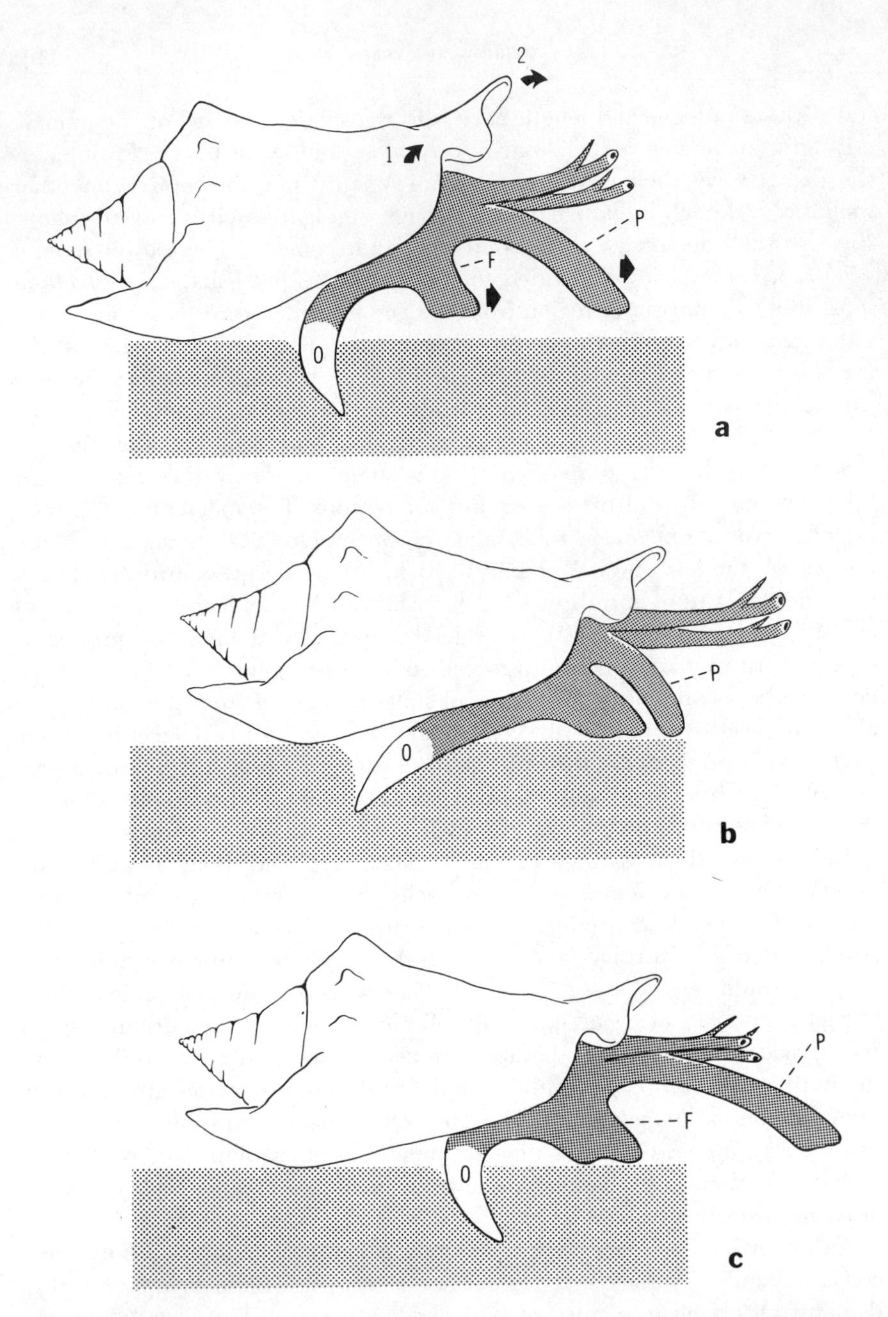

FIGURE 8. Stages in the locomotion of conchs: (*a*) hunching the shell forward; (*b*) forward movement of the proboscis; (*c*) pulling forward of the propodium.

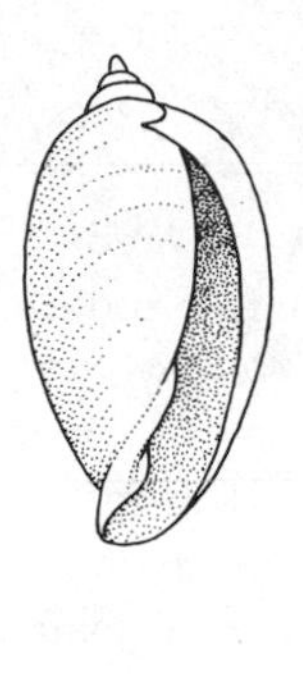
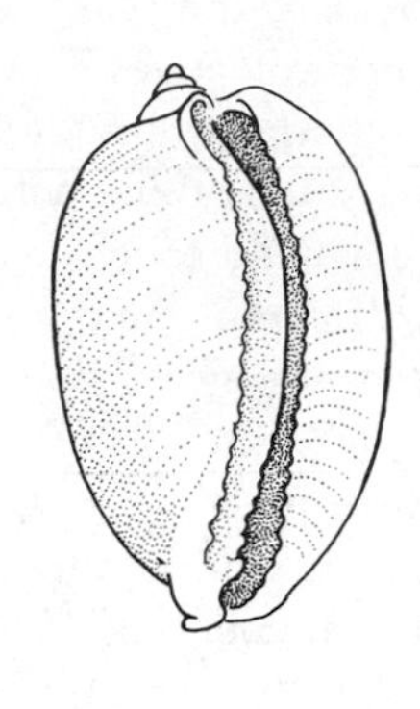
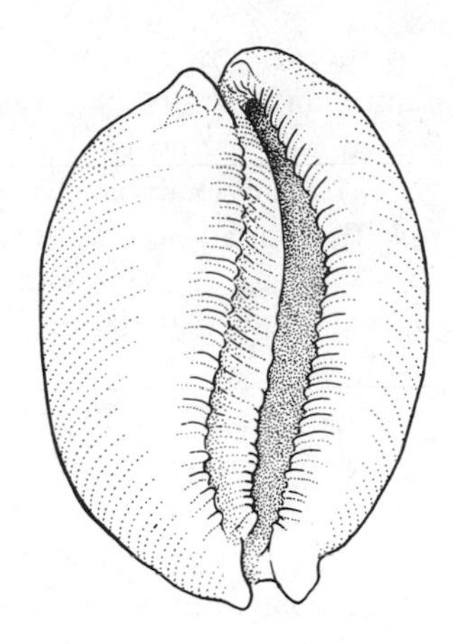

FIGURE 9. Shell coiling in young (left), subadult (middle), and adult (right) cowries showing development of thickened lip and apertural teeth.

the dorsal margin. The closely spaced black spots indicate the degree of mantle extension. At the anterior end (partly in shadow) can be seen the slender tentacles with small black eyespot near the shell edge. The difference in mantle and shell color is startling in *Cypraea cribraria* (Plate 6, center and lower right). The shell is brown with large white spots, while the mantle is bright red and can be almost fully (center) or barely (lower) expanded. The mantle surface has many short projections or *papillae* (see upper margin). These are thought to function in respiration. At the anterior end, protruding from the white rim of the shell, is a short trumpet-shaped *siphon*. This rolled edge of the mantle cavity serves to funnel water into the mantle cavity and is lined with sensors. The water exits along the right side of the head in typical mesogastropod fashion.

The papillae on the mantle differ among species. In *Cypraea saulae nugata* (Plate 6, upper left), there are only a few yellow *dendritic* (branched) papillae scattered over the surface of the light orange, rather translucent mantle. By contrast, *Cypraea limacina facifer* (Plate 7, upper left) has a dense thicket of sometimes trilobed, tubular, long papillae. The rather long siphon and one tentacle are clearly shown near the right margin. In *Cypraea subviridis* (Plate 7, upper right) the brown foot and mantle contrast with the lighter shell. The white mass to the left of the animal is composed of eggs laid by the cowry. The rock is covered with a reddish encrusting organism and has been turned over from its natural position for photography. The eggs are laid in clusters in crevices or under coral boulders.

Although observational detail is lacking, the reason for the difference in mantle and shell color is thought to be protective. The mantle can be withdrawn quite quickly. If a predator bumped into an expanded *Cypraea cribraria,* for example, he would touch something soft, plump, and bright red. If the mantle withdraws before a second touch or a grab by the predator, it will see or touch something hard, smooth, and brown with white spots. In theory this will confuse the predator enough so that it will look elsewhere for dinner.

Several families of the Mesogastropoda, the Cymatiidae, Cassidae, Bursidae, Tonnidae, and Ficidae, have reached large size and experimented with carnivorous diets, usually on echinoderms. The nearly 2-foot-long *Charonia tritonis* is a predator on the "Crown-of-Thorns" starfish, whose population explosion is devastating some Pacific coral reefs. In these higher mesogastropods, the siphon is somewhat elongated with the anterior shell prolonged into a siphonal canal, as in *Ficus* (Plate 4, second from top), which also has lost its operculum completely, an unusual feature in a mesogastropod. Excellent illustrations of mesogastropod variety and a concise synopsis of their poorly known biology is given in Wilson and Gillett's *Australian Shells*.

The basic reference on prosobranchs and the finest book on mollusks issued in this century is Fretter and Graham's *British Prosobranch Molluscs*. This is more than highly recommended.

From the bewildering variety of "mesogastropods" came three major groups: the pulmonates, which dominate the land snail fauna; the opisthobranchs, which show exuberant experiments in shell reduction with specializations in habitats and feeding; and the stenoglossan prosobranchs, which are scavengers and poison bearing hunters on other snails and even fish. The marine stenoglossans and opisthobranchs are reviewed next, while the nonmarine taxa are deferred until subsequent chapters.

VII
Sophisticated Hunters

Throughout the animal kingdom there is a tendency for those organisms practicing "passive" resistance to be heavily armored, to have a color that blends into a natural background, and/or to behave in as inconspicuous a fashion as possible. In contrast, there are organisms equipped with an aggressive means of defense such as a poison sting, as is found in a bee or wasp, or which simply taste horrible. Frequently they are very brightly colored, have reduced their defensive armor, and move about fearlessly.

Both archaeogastropods and generalized mesogastropods tend to clamp onto a rock (limpets) or retreat into their shell behind a stout operculum as a means of defense, but the higher mesogastropods, stenoglossans, and opisthobranchs have become more aggressive and highly colored predators, although going about this in different ways (Plates 5, 7, 8, and 9). The opisthobranchs have remained small in size, tend to lose any trace of a shell, became highly specialized in feeding on small organisms, and are among the most brightly colored (Plate 9) and most camouflaged of all mollusks. The stenoglossan prosobranchs include the largest living snail, became very active boring predators, scavengers, or poison gland equipped hunters, retained a solid shell, have the operculum reduced to a small remnant, and hunt relatively large prey. There are common patterns of change within each group and highly diverse specializations that can only be sampled in the space available here.

STENOGLOSSAN SNAILS

Before a predator can eat, it must locate, pursue, and capture its prey. Those predators feeding on attached barnacles or epifaunal clams such as *Arca* or *Mytilus* have a very different problem from those hunting an actively burrowing clam or crawling worm. The former can sense a mass of attached individuals, move over at comparative leisure, and munch away as their appetite dictates. But to locate and catch a single moving individual requires better "detecting" equipment and specialized catching behavior.

The mesogastropod mantle cavity arrangement allows a continuous flow of water in the left side and out the right. Critical sampling of the incoming water for chemical clues becomes much more feasible than in archaeogastropods where the currents inevitably will swirl around a bit. One of the main changes in the stenoglossan snails is the development of a long siphon from the mantle edge. In the mesogastropod cowries, the siphon is a short to medium length tube (Plates 6 and 7). *Ficus* has the anterior of the shell tapered and extended moderately in front of the head (Plate 4, second from top). Only the siphon tip protrudes from the shell. In the stenoglossan volutes (Plates 7 and 8) and *Conus* (Plate 4, bottom) the siphon is a long, highly flexible tube that can be extended above or well in front of the snail's body. Many species have the siphon as a closed tube, but most have it open on the bottom (Plate 8, top). The inside of the siphon is lined with chemoreceptors that test for minute traces of chemicals from a potential victim. Thus the siphon is used to "smell out" and track down a meal.

Although many groups have a freely movable siphon, in others the shell is modified to provide protection for it. Perhaps the ultimate extension is seen in *Murex pecten* and *Murex acanthostephes* (Plate 5, top two). The snail's body is relatively small, while the shell has a very long, narrow, spine-covered tube extending far in front of the head. This encloses a siphon, which obviously can be poked into places and sample water far in front of the snail's head.

Once the victim is located, then it must be captured and entered. Here stenoglossans use a variety of techniques. The muricids generally bore holes in the shell of their victim, using a nonacidic secretion from a foot gland to soften the victim's shell. Species of the Atlantic Ocean *Busycon* will batter a clam with the lip of its own shell until a big enough hole has been chipped away in the clam shell for *Busycon* to feed. The West Central American fasciolariid *Opeatostoma pseudodon* (= *Leucozonia cingulata*) has a spine on the anterior shell lip that apparently is used to wedge open the shell of a clam. Other fasciolariids use the siphonal canal to hold open the victim's shell, as do various muricids and buccinids. Toxoglossan species, Superfamily Conacea, have, at least in many species, a poison gland as part of the feeding apparatus. They use a single radular tooth that is shaped like a hypodermic needle, literally spearing their food, and then injecting poison that paralyzes their prey.

Many others are scavengers instead of active predators. The common *Melongena corona* of Florida (Fig. 1) depends on a highly extensible proboscis. A specimen with a 3 inch shell can easily reach clam meat at the bottom of a 6 inch test tube. Mitrids will swarm around the pipes from outhouses, although normally they are carrion eaters. Nassariids on mud flats and olivids in sandy bottoms are numerous and active, with extremely long and flexible siphons. "Baiting" all of these snails with pieces of fish or mollusks is a very effective collecting technique.

In addition to the more highly developed siphon, stenoglossans usually have only three radular teeth in each transverse row (see Chapter IX, Fig. 5), instead of the seven found in most mesogastropods. In the toxoglossans, a bundle of harpoonlike teeth sit in the radular sac, but only one is used at a time, which cannot be reused subsequently. In typical muricids, such as *Murex acanthostephes* (Plate 5, top) a horny operculum (brown flat disk to the left of the body stalk) can close the aperture completely. Some of the smaller species, such as *Drupina grossularia* (Plate 5, lower two) have projections from the apertural edge (bottom) that narrow the opening, and the operculum is correspondingly reduced in size. Those carnivores that are equipped with poison glands (Conidae, "Turridae," Terebridae) have the operculum reduced to a nonfunctional sliver on the foot.

Although the basic anatomies of the stenoglossans are quite comparable to each other and basically show only the few trends in structure outlined above, they are quite diverse in shell shape and ornamentation. Previous discussion of the gastropods has focused on the soft anatomy and progressive evolution, but here an excursion into conchology is warranted. Figure 1 diagrams the basic parts of a gastropod shell, using the fasciolariid *Fusinus* and the melongenid *Melongena* as examples. Both are simple spiral shells, coiled in regularly increasing width around an imaginary line called the *shell axis* (1 in Fig. 1). Each complete loop of the shell around the axis is called a whorl (spiral arrow). The early part of the shell, carried pointing posteriorly in life, is called the *nuclear* or *apical* whorls (2). These form in the egg capsule of stenoglossans, or are the first formed part of the shell in those species with pelagic larvae. The opening of the shell is called the

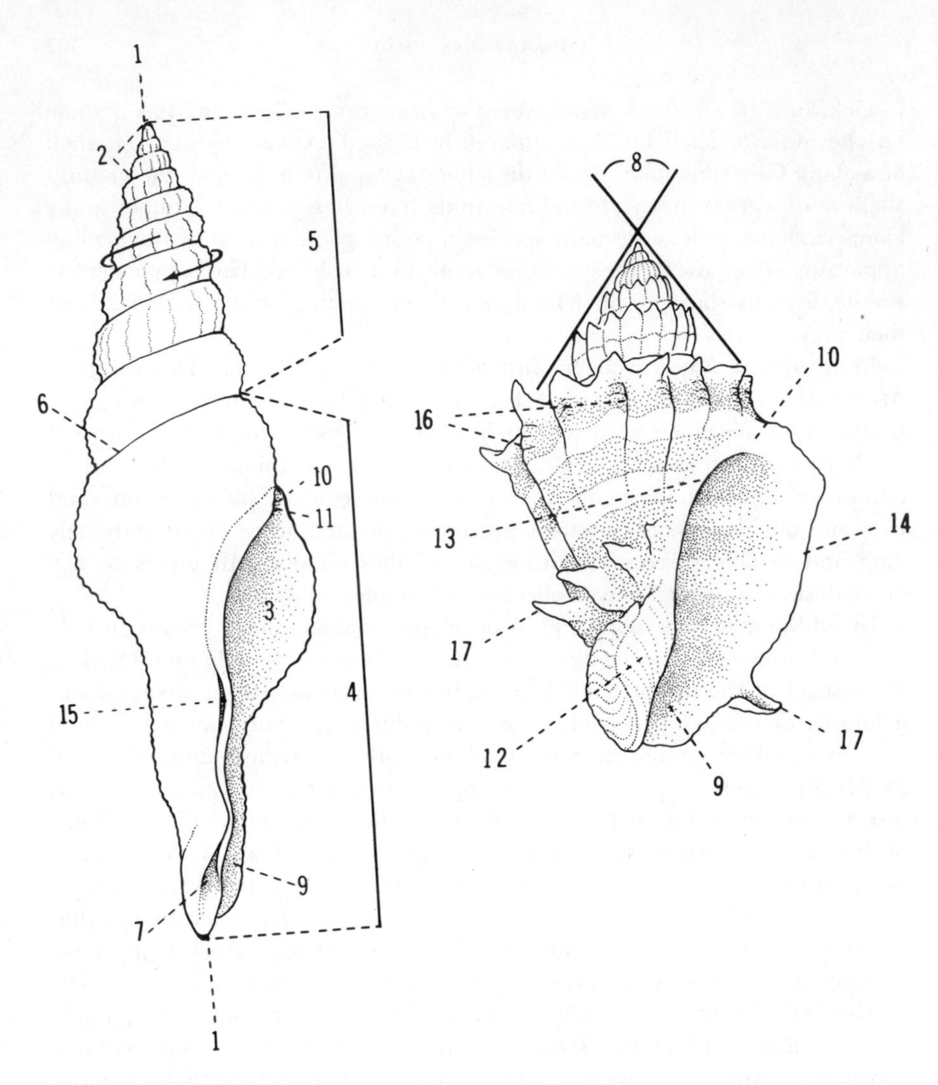

FIGURE 1. Parts of the gastropod shell, based on *Fusinus* (left) and *Melongena* (right). Structures shown are shell axis (1), nuclear or apical whorls (2), aperture (3), body whorl (4), spire (5), suture (6), umbilical slit (7), spire angle (8), siphonal canal (9), anal canal (10), parietal ridge (11), columellar lip (12), parietal or inner lip (13), palatal or outer lip (14), columellar callus (15), ridges (16), and spines (17).

aperture (3), and the whorls leading up to the aperture are divided by custom into two regions—the *body whorl* (4) for the last complete whorl before the aperture, and the *spire* (5), which includes all the other whorls. The line marking the outer edge where a lower whorl (further down the spire) is coiled against the previous one is called the *suture* (6). At the anterior end of the shell there may be a narrow to widely open space called the *umbilicus* (7). The rate at which the whorls widen on the spire is measured by the *spire angle* (8). If the whorls widen rapidly and coil around each other tightly so that there is a wide spire angle, then the umbilicus may be as widely open as it is in many land snails. If the coiling is very "tight," as in the figured species, then the umbilicus may be only a narrow slit (*Fusinus*) or even closed (*Melongena*).

Just as the inside shell features of a clam tell something about the animal that made it, so the structures of the gastropod aperture provide data about the snail. Significant features include whether the *siphonal canal* (9) is long or short, whether the anal canal (10) is well developed or just a notch marked by a *parietal ridge* (11), and whether there are any columellar folds or plicae on the *columellar lip* (12), and structures on the *parietal* or *inner lip* (13) and *palatal* or *outer lip* (14). Often the columellar or parietal lips will have a raised *callus* (15).

Shell growth is frequently not regular, but instead occurs in spurts. During the pauses between spurts of growth, complex ridges (16) or spines (17) can be formed. At times these can be quite regular and spaced close together (*Melongena*), or more widely spaced as the spines on *Murex pecten* (Plate 5, second from top) and the anal siphons formed around the whorls of another muricid, *Murex cornutus* (Fig. 2, left). In contrast, the few ridges on the cymatiid *Cymatium rubeculum* (Fig. 2, right) are not formed at either regular or equal intervals. In a very few species, such as another cymatiid, *Biplex perca* (Fig. 3), flaring varices (singular, varex) are formed at half-whorl intervals, giving a "winged" appearance to the shell.

There are also considerable differences in the patterns of coiling shown by different species. The toxoglossan families Conidae and Terebridae have very similar anatomy, if allowances are made for their different shell coiling pattern (Fig. 4). *Conus marmoreus* (upper shell) has the whorls winding around each other with the result that it has a very short spire compared with body whorl length and an extremely wide (140°) spire angle. Each whorl is long and tapering, but the coiling pattern is almost in a single plane. In contrast, *Terebra areolata* (lower shell) has the whorls barely connected to the bottom of the previous one. The shell is drawn out into a long cylinder shaped like a spike or augur (hence the common name of "augur shells"). As a result the spire is very long, the body whorl quite short in comparison,

FIGURE 2. *Murex* (*Bolinus*) *cornutus* (left), a muricid snail from West Africa, and *Cymatium rubeculum* (right), a cymatiid from the Great Barrier Reef off Australia. *Cymatium* is 2½ inches long. Courtesy Field Museum of Natural History, Chicago.

and the shell has a very narrow (15°) spire angle. In this species of *Conus* the spire is about one-twelfth the body whorl length, while in the *Terebra* the same ratio is 3.5. Other species in these families show somewhat intermediate shell proportions, but even the most similar species from each family are noticeably different in spire height.

Shell variations are almost infinite in number, and most books on shells present extensive data on this part of the animal. Here the emphasis is on evolution and biology rather than shell structure. Before leaving the stenoglossans, I will present two more aspects of variability. Closely related species can live together only if they avoid direct competition with each other, either for food or some other aspect of the niche. If two closely related species do compete directly, usually there will be areas where each one dominates in numbers, and a shifting, variably wide "zone of contact" where neither species can completely replace the other. If the mollusks involved have larvae that spend days or weeks floating in the plankton, then the species will get widely distributed. Competition avoidance between these species usually means clear-cut specialization on different foods or picking different microhabitats in the same area. Narrow zones of contact cannot be maintained

FIGURE 3. Apertural and top views of the Japanese deep water cymatiid, *Biplex perca*, collected at 50 fathoms off the Kii Peninsula of Japan. Shell length 2½ inches.

FIGURE 4. *Conus marmoreus* (top) from the Philippine Islands and *Terebra areolata* (bottom) from New Guinea. The *Terebra* is 4 inches long.

with planktonic larvae. Many species fasten egg capsules to a rock or lay them on the sand. The young emerge from the egg capsules as crawling young. Dispersal in these species is limited to the distance that a snail can wander during its lifetime or the results of accidents. In these situations, sharply delimited ranges and narrow zones of contact are quite possible.

Volutes (Family Volutidae) and cones (Family Conidae) illustrate the two types. Volutes apparently only lay eggs that hatch into crawling young, while reproduction in cones ranges from planktonic larvae to crawling young types.

The coast of Australia has perhaps the greatest number of volutes found anywhere in the world. Many of these are highly restricted in their distribution, and differences between species of the same genus can be striking. The little that is known about their feeding habits suggests that volutes are scavengers and opportunistic predators. They do not show any great specializations in food. The living animals are even more spectacular than the shells (Plates 7 and 8). There is no correlation between shell and animal color. Within the same genus there can be enormous differences. *Amoria maculata* (Plate 8, upper two) occurs in shallow waters off central and northern Queensland. The radiating white stripes and ringed siphon on its red body are striking, but pale in comparison with the brilliance of *Amoria canaliculata* (Plate 8, lower two). This species, trawled in deep waters off central Queensland, has a shell with reddish-brown blotches and wavy lines on a white background, while the animal has patterned blotches of brown, white, and yellow. The contrast between shell and animal is less in *Cymbiolacca pulchra* (Plate 7, center left), while the young *Melo amphora* (Plate 7, lower left) is positively dowdy in comparison. The detailed pictures show that the siphon is not a completely closed tube and give a good idea of the head and tentacle structure in volutes. The *Melo* photograph also shows the bulbous nuclear whorls (white in color) which are characteristic of many mollusks that hatch in a crawling stage.

Volutes have limited distribution and, as mentioned above, show no evidence of food specialization. *Conus* shows wide species distributions and great food specialization. Primarily due to the investigations of Alan Kohn, considerable information is available on the feeding habits and general ecology of *Conus* species. All cones studied to date are active predators, feeding on animals in one of four categories—polychaete worms (Phylum Annelida), members of obscure phyla, other mollusks, or small fish whose body length may exceed that of the snail's shell. While many fish eat mollusks, the cones are the only snails known to strike back. The fish feeders are quite spectacular in action, as described in the following, which is based on several filmed sequences: The snail lies buried in the sand with the tip of its siphon extended above the surface. Many coral reef fishes have the habit of hovering just above the surface of the sand. If a fish pauses above a hungry species such as *Conus striatus,* the cone's proboscis will be extended above the surface and move upward until it touches the soft underbelly of the fish. Instantly one of the big radular teeth will be pushed into the fish's belly and a very potent neurotoxin injected. The radular tooth is not released by the cone, but held tightly by muscles at the tip of its proboscis. The literally harpooned fish will wriggle for a few seconds, and then becomes paralyzed. Finally *Conus* rears up out of the sand and swallows the fish whole, sometimes folding it in half in the process.

Feeding differs slightly in those *Conus* that eat mollusks. Here the radular tooth is shoved into the victim, but not held by the cone. Sometimes several radular teeth may be injected one after the other. Since the mollusks cannot crawl away as quickly as the fish can swim, holding onto the dying victim is unnecessary. If the victim is a sea slug or an opisthobranch with thin shell, then shell and animal will be swallowed by the cone. If the victim has a large and heavy shell, then the *Conus* will hold its mouth against the shell aperture of the victim. Possibly the poison serves to loosen the columellar muscle (the only point at which a snail is attached to its shell) so that the *Conus* can suck its dinner from the shell. *Conus marmoreus* (Fig. 4) is one of the mollusk eaters. The majority of cones feed on marine worms. Here a single radular tooth is ejected, which may or may not be held by the proboscis until the worm stops squirming.

Within any particular area, worm-, mollusk-, and fish-eating *Conus* can coexist together. More species can be packed into an area by specialization in habitat. In Hawaii there is a logical division of the habitat into intertidal marine benches and subtidal coral reefs. Alan Kohn found that species of *Conus* differed in habitat preference and abundance within microhabitats of these broad areas. Of 22 *Conus* species studied by Kohn, 11 feed only on worms, three on mollusks, three on fishes, and the remaining five feed on both worms and some other marine animals. Their diversity is maintained by food specialization.

Any live *Conus* should be treated with respect by a collector. Several human deaths have been caused by careless handling of *Conus textile* and *Conus geographus*. Potentially all fish-eating *Conus* are capable of inflicting severe injury, if not death. Since the vast majority of the 500 living species of *Conus* have not yet been investigated as to feeding habits, and there are no shell distinctions between fish-eating and worm-eating *Conus,* common sense requires cautious handling. The characteristic shell form is not duplicated by other marine mollusks, and the living animal with its inquisitive siphon and usually narrow appearing foot is easily identifiable. *Conus ammiralis* (Plate 4, bottom) is typical in appearance. Like most species, its food habits are unknown.

These then are the stenoglossan prosobranchs, medium to large sized scavengers and predators in shallow to deep ocean waters. Reduction in the number of teeth in each row on the radula, frequent loss or great reduction in opercular size, development of a long and highly efficient siphon, plus the development in the Conacea of a poison apparatus mark the major advances of the group. They are the dominant sea snails of today. Except for a genus of Buccinidae, *Clea,* found in the rivers of Borneo, and the $\frac{1}{3}$ inch long marginellid, *Rivomarginella,* from Thailand lakes and rivers, none have migrated into freshwater. Because of their large size, reduced operculum, and dependence on sampling water currents to locate food, migration onto land was an option closed to them.

OPISTHOBRANCHS

Nobody knows how many species of opisthobranchs are alive today, since the work of sorting out species from color variations in sea slugs has barely started. An educated guess might be that some 2500 species have been described well enough so that they can be identified. Anybody who attempts to collect sea slugs for study finds that a good percentage of the species do not fit any of the previous descriptions. They can be presumed to be unknown, often for reasons of simple neglect and oversight.

The now classic example of this concerns the "bivalved gastropods" of the Family Juliidae. For three-quarters of a century some $\frac{1}{20}$ to $\frac{1}{3}$ inch shells, ranging in age from Eocene fossils to recent species, were classified as clams. In the late 1950s a Japanese zoologist, Siro Kawaguti, brought back some algae of the genus *Caulerpa* to his laboratory. To his intense surprise, a few of the grapelike objects on the algae, which seemed to be merely part of the alga itself, began to crawl. The rest were part of the *Caulerpa*, but the snail, *Berthelinia limax* (Fig. 5), is now famous. *Berthelinia* has the

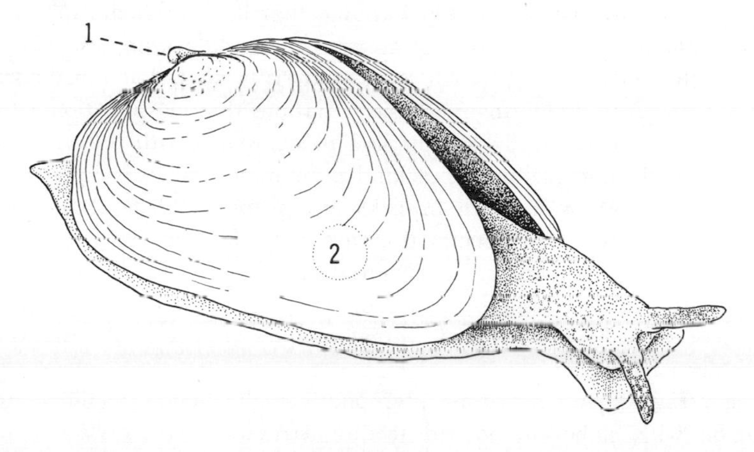

FIGURE 5. *Berthelinia limax,* a bivalved snail. Actual shell length is $\frac{1}{8}$ inch. The original coiled shell (1) and adductor muscle (2) are shown. Based on original water color drawings of top, bottom, and side views of living animal by Siro Kawaguti.

anatomy of a sacoglossan opisthobranch, but has evolved a two-valved shell even though it starts out in life as a typical larva with a single coiled shell (1). A few days after hatching, projections from each side of the larval shell extend downward, a larval retractor muscle shifts in position to form a single adductor muscle (2) that is visible through the shell, the dorsal margin of the shell becomes joined by an elastic ligament, and weak hinge teeth develop on the valve margins. The similarity to bivalves is an example of convergent evolution, the bivalve shells having evolved independently.

Berthelinia lives among the grapelike "leaf" clusters of *Caulerpa* and has the same feeding habits as other members of the opisthobranch Order Sacoglossa. There is only a single radular tooth in each row which is used to pierce the algal cells, the contents of which are then sucked out by the mouth. Other members of the Sacoglossa include shell-bearing and sluglike species, but all seem restricted to a few types of algae. Although living Juliidae were not reported until 1959, within six years they had been collected in nearly every ocean where *Caulerpa* exists. This indicates how much work still remains to be done in simply inventorying the mollusks of the world.

Opisthobranchs evolved from the mesogastropods, but how many groups evolved into opisthobranchs is uncertain. Many studies have been made on the anatomy of groups that combine prosobranch and opisthobranch characteristics, such as the Pyramidellidae and Homalogyridae. At the same time there are families such as the Onchidiidae that have been purported to be opisthobranchs, although showing stronger pulmonate affinities. The situation is highly confused and much more study is needed. The opisthobranchs show such an incredible range of basic variations that classification into perhaps 11 orders is generally accepted. This compares with only two orders of prosobranchs and five orders of land pulmonates. What is equally significant is that the orders of opisthobranchs, except for the Order Nudibranchia, usually have only a few species that are very different from each other. There have been exuberant experimentation with sharply different ways of living, patterns of structures, and food specialization, but only the nudibranchs seem to have developed a great number of species. A few examples of this diversity follow.

The pelagic gymnosomatous pteropods, small shelless members of the plankton with elaborate hooked feeding structures, are only one of the groups that swim. In many opisthobranch families there are lateral extensions of the body called *parapodia.* By flapping these many species can swim awkwardly. A few hold the parapodia to form a primitive jet. Others, such as the nudibranch *Hexabranchus* (Plate 9, bottom), swim by undulations of the whole body. This dinner-plate size swimmer has been compared to a ballet dancer in grace as it moves by undulating the sides of the body. A single undulation takes about 4 seconds. This is one of the most beautiful sights in tropical waters.

Also shown in Plate 9 are a species of *Phyllidia* (top figure), and two dorid nudibranchs with egg masses. This is only a hint at the variety of nudibranchs, the most beautiful but least studied group of mollusks. Among their unusual features is the ability of some eolidacean species to defend themselves by taking stinging cells (nematocysts) from the bodies of small coelenterates. The cells do not discharge when eaten, but pass through the wall of the nudibranch's intestine, and are moved to the tips of the cerata (tubular projections) on the sea slug's back. The stinging cells are then oriented pointing outward and will be triggered if touched by another animal. This taking of the defense mechanism from a species in one phylum and making it a functional part of the body in a species of another phylum is, to my knowledge, a unique phenomenon. It has been reported, however, for a number of eolidacean sea slugs.

Equally unique is the ability of some sacoglossan opisthobranchs to incorporate the photosynthetic organs (chloroplasts) from their algal food into their own digestive gland cells. The apparently unharmed chloroplasts can continue to function in photosynthesis (producing organic matter) for as long as six weeks in one species, or as little as one day in another. The chloroplast formed by the alga is "taken over" by the snail who in "consuming" a "producer" has temporarily become a partial producer itself. It continues to gain organic matter by means of its victim's still functioning body part.

Only a few adult opisthobranchs have shells, but all larval forms have shells. The one clear trend in evolution is toward loss of the shell. In this process, the animal becomes "detorted," that is, the mantle cavity twists and is shifted to the right posterior portion of the body. Since opisthobranchs are descended from mesogastropods, only one gill was present in the ancestral stock. This is the positional "left" pallial gill that had rotated from the "right" position at the posterior in the ancestral mollusk. In opisthobranchs this returns to a right side position, although in the nudibranchs and some other groups it is completely lost. When this happens the openings that normally empty into the mantle cavity instead open directly to the exterior through the body wall.

Opisthobranchs are exclusively marine, except for about a half dozen species of acochlidiaceans taken in freshwater streams in Indonesia and the Palau Islands that resemble nudibranchs. Highly selective and sophisticated in their food habits, brilliantly colored as a warning to predators, or perfectly camouflaged against their background, they are fascinating but little studied creatures. It is indicative of their obscurity that only the review by Libbie Hyman in *Mollusca I* can be suggested as a major review source, although the treatment in Morton's *Molluscs* is helpful.

VIII
Vanishing Freshwater

Organisms moving from the oceans into freshwater face a new set of problems. Of immediate concern is the question of regulating their body fluids. Except for the vertebrates, most marine organisms have the concentration of salts in their body essentially the same as that of the surrounding seawater. The exact ratio between particular salts will vary, but the overall balance is very close to the 3.5% average salts content for the oceans. In contrast, bodies of freshwater average about 0.01% salts, except for salt lakes or other special cases. Where water flows from land into the sea there will be a mixture and zone of transition from the "salt-free" freshwater to the "briny ocean." Freshwater organisms will tend to have water flow into their bodies through gills or skin which will upset their salt balance by diluting their body fluids. To correct this freshwater organisms must excrete large quantities of water that is salt free. Often the surface areas of the body through which water can enter the tissues will be reduced in size.

A second major problem concerns reproduction. In the oceans, a larval swimming stage disperses members of a species over a wide area. Some larvae of shore dwelling gastropods have been netted in the mid-Atlantic, since the currents carry the young for great distances. But in freshwater such a swimming stage would at best only carry the young downstream and at worst it might take them out to sea. So great changes in egg laying and larval life have been necessary.

A third major problem is that most bodies of freshwater are isolated from each other, whereas all of the oceans are interconnected and thus continuous. A single drainage, like the Ohio River, will be continuous from its headwaters to its junction with the Mississippi, at least before man-made dams interrupted and controlled its flow, but the headwaters of the Tennessee and Ohio rivers are separated by land areas and quite isolated from each other. The Hudson and Ohio rivers are isolated from each other and completely separate from the Columbia River of Washington, much less the Ob of Siberia, Amazon of South America, or the Nile of Africa. In some local areas, particularly those with glacial moraines (debris piles left behind by the slow retreat of glaciers), there will be ponds, small lakes, swamps, marshes, and temporary pools in the rainy season. Wherever man has settled, there are watering troughs for cattle or horses, roadside ditches, and literally hundreds of other such spots from which water does not drain. They will hold water for at least part of the year. All such sites are potential places for freshwater organisms to colonize, *if they can get to them.* So solving the problem of "overland" dispersal has been a major key to successful freshwater living.

The fourth major problem is really an extension of the third. Most bodies of standing freshwater are quite temporary. Except for a few ancient lakes, Titicaca in South America, the African Rift Lakes (Victoria, Nyasa, Tanganyika), Ohrid in Yugoslavia, and Baikal in Siberia, lakes result from a temporary catastrophe and have a "life span" of less than 50,000 years. Even the oldest lake, Ohrid, may be only 200,000 years old. An earthquake, landslide, or man-made dam interferes with the flow of a stream or river. This barrier backs up the water, spreading it out over a greater area and forming an impoundment of still or very slow moving water. All water running off land surfaces carries particles of dirt in suspension and leaves or other debris. When the water runs fast, the particles are carried along. As the water slows down, more and more of the particles settle out and add to the bottom sediments. Water flowing into a lake carries such particles. When slowed, they are dropped. In time the lake or dam site will fill up with sediment and become first a marsh or swamp, then eventually dry land. Drainage *systems,* however, are more permanent. Once mountains are elevated, runoff water on their slopes will continue to move toward lower elevations until the mountains themselves are eroded away. Channels of individual streams or rivers are subject to great changes, but drainage basins may maintain themselves as a unit for long periods of time.

The limitations and restrictions just described have had a profound effect on freshwater life. Comparatively few groups of animals have colonized freshwaters. Among those there are great differences in the extent and pattern of diversity. Crustaceans, insects, annelid worms, snails, clams, water mites, nematodes, and flatworms are the most abundant, with only scattered representatives from other animal phyla. Those that did diversify were adapted to freshwaters as they existed prior to large scale agriculture and high concentrations of human populations.

Particularly in regard to organisms inhabiting flowing waters, man's agriculture and habits are incompatible with the needs of the organisms. First of all, the amount of dirt washed away from a forest, grassy meadow, or well-tended lawn is a tiny fraction of the quantity washed away from a plowed field, cleared lumbering area, or unseeded bare soil around a new house. The silt burden produced by man smothers many organisms and exterminates aquatic life from many streams. Secondly, running water is an irresistible temptation for us to dump wastes of all kinds—garbage, industrial, mining leftovers, sanitary. Western industrial civilization has been built on the thesis of "I can dump whatever I want into flowing water and to hell with anybody downstream." Some chemicals directly poison freshwater life, as was exemplified by the hundreds of "million plus" fish kills of the 1950s and 1960s. These were caused by pesticide runoffs or chemical discharges into streams. Biological wastes, in reasonable quantities, can be used as food by stream dwellers, but above a certain level they overwhelm the stream life by depleting the oxygen level and produce "aquatic deserts." Third, there is a vicious cycle of intensified agriculture producing increased runoff during rains and leading to higher human population density. The increased runoff means that downstream flooding becomes worse. The increased population means that more and more people will try to settle on flood plains (lowlands over which water spreads during floods). People thus move further onto the "floodable areas" at the same time their activities are enlarging the area subject to flooding. This leads to the "demand" for a dam to hold floodwaters, which changes a flowing stream that occasionally spreads over a flood plain into a lake that permanently floods one area (until it silts up in a few years). Dams are frequently justified because they produce a "recreational lake," serve as a water "reservoir," or generate electric power. At times they are of great value, but the reasonable construction of dams in North America has long since been exceeded.

In terms of animal extinction, the dams of Eastern North America are producing an unparalleled disaster. Probably 40 to 50% of all the freshwater molluscan species in the world lived in Eastern North America, most of them occurring nowhere else. The majority of these evolved in the "shoal" or "riffle" areas of streams, fast moving shallow water over a sandy or gravel bottom. In 1968 it was estimated that 40% of the unionid clam species in the Ohio River drainage were extinct or in immediate danger of extinction. In rivers such as the Coosa of Alabama, the degree of actual extinction is far higher. Some stretches of these rivers have had their fauna extinguished by silting or pollution by city and industrial wastes, but the major factor has been dam construction. This changes a riffle habitat with high oxygen and plentiful food into relatively stagnant water with low oxygen content and little food. Some species become extinct behind the new dam immediately, others are unable to reproduce and thus extinction is "postponed" until the last survivor dies. A symposium on North American rare and endangered mollusks, sponsored by the American Malacological Union in 1968 and published in 1970, concluded that

> . . . more than 400 native species of mollusks are in imminent danger of extinction. At least 1000 others will soon be endangered if present trends continue.

Included in these numbers are 40% of the unionid clam species known from the Ohio River drainage and more than 140 species of freshwater clams and snails from the Southeastern United States. This mass loss of species is the result of habitat alterations made by man during the last 100 years. David Stansbery concluded his contribution to the symposium with the statements:

> We are left with the inescapable conclusion that we are gradually destroying nearly a thousand endemic species of freshwater mollusks. This fauna was millions of years in coming into being and is in the process of being eliminated in only a century or two.

The situation is best documented for Eastern North America, but it can be anticipated that as the spread of flood control and dam formation intensifies along the major drainage systems of Southeast Asia, South China, Southern South America, and parts of Africa, similar mass extinctions of freshwater mollusks will occur. The numbers involved will not be as dramatic, since the diversity of freshwater life is not as great, but the phenomenon will be the same.

This problem affects those species with low dispersal powers which inhabit larger bodies of water. The species able to reach temporary ponds and ditches are relatively unaffected. They are also fewer in numbers of species and are usually far less interesting to biologists. Both clams and snails are abundant in freshwaters. A group by group introduction to these now follows.

FRESHWATER CLAMS

Four groups of clams invaded freshwaters. By far the most conspicuous and diverse are the "unionids," 1 to 10 inches in length, comprising about 1000 species divided among several families, including one group (Family Aetheriidae) whose shells become cemented to the bottom by one valve. These are known as freshwater oysters (Fig. 1) and occur in scattered areas of South America, Africa, and India. Next in abundance are the fingernail clams of the Family Sphaeriidae. These are $\frac{1}{8}$ to $\frac{1}{2}$ inch in diameter bottom dwellers, often extremely abundant, and form a substantial portion of the food taken by many fish. Their young are brooded inside the gills of the

FIGURE 1. A freshwater oyster. *Acostaea,* from the Magdalena River of Columbia, South America. Length $3\frac{1}{2}$ inches.

adults and held there until they reach crawling size. Depending on the species, from 2 to 20 young may be present at any one time. Sphaeriids are found in most permanent bodies of freshwater, where they are occasionally accidentally transported by flying insects or on the feet of water birds.

Members of the Family Corbiculidae were originally absent from North America, although abundant on other continents. Some time in the late 1930s an Oriental species, *Corbicula manilensis,* was accidentally introduced into the United States. It has spread rapidly in the Western States and through much of the Tennessee River system, and will spread much further. At times dead shells can clog irrigation canals several feet deep. The fertilized eggs are brooded inside the gills of the parent and released as swimming young. During a short stage they float in the current, thus being widely dispersed, but soon settle to the bottom, sometimes briefly attached by a byssus, and eventually become burrowing filter feeders. If the floating larvae attach to the sides of water inlet pipes, they can easily grow and clog the pipes completely. In Europe and Asia, *Dreissena polymorpha,* a byssate veneroid clam with a swimming larval stage, has apparently rather recently made the transition from brackish to freshwater situations. It is causing problems similar to those resulting from the spread of *Corbicula.*

The "unionids" are not only the largest in the number of species but biologically perhaps the most interesting. As is typical in the study of clams, their classification is hotly debated between those who depend mostly on the hinge structure and those who study the gills and reproductive systems. Something of a general consensus is emerging. There are distinctive larval types. When this is combined with variations in shell structure which correlate with major geographic areas, a logical and simple classification results:

SUPERFAMILY UNIONACEA

FAMILY MARGARITIFERIDAE (North America and Eurasia)
FAMILY AMBLEMIDAE (United States and Canada)
FAMILY HYRIIDAE (South America and Australasia)
FAMILY UNIONIDAE (worldwide)

SUPERFAMILY MUTELACEA

FAMILY MUTELIDAE (Africa)
FAMILY MYCETOPODIDAE (South America)

The split into two superfamilies is based on differences in how the clam larva functions. Unionids solved the problem of dispersal in a unique way, that is, by spending part of its larval stage as a parasite, usually on a fish.

In species with separate sexes, sperm are released into the water, frequently with the male individuals facing downstream. The females are facing upstream and the sperm is pumped into the inhalent siphon and passed into the gills. Eggs have been released into the "water tubes" inside the gill of the female, where fertilization takes place. The fertilized eggs are brooded inside the gills until they grow into a stage called a "glochidium" in the Unionacea (Fig. 2*e*). At this point the larvae are expelled from the exhalent siphon and drop to the bottom. They lie with their valves opening

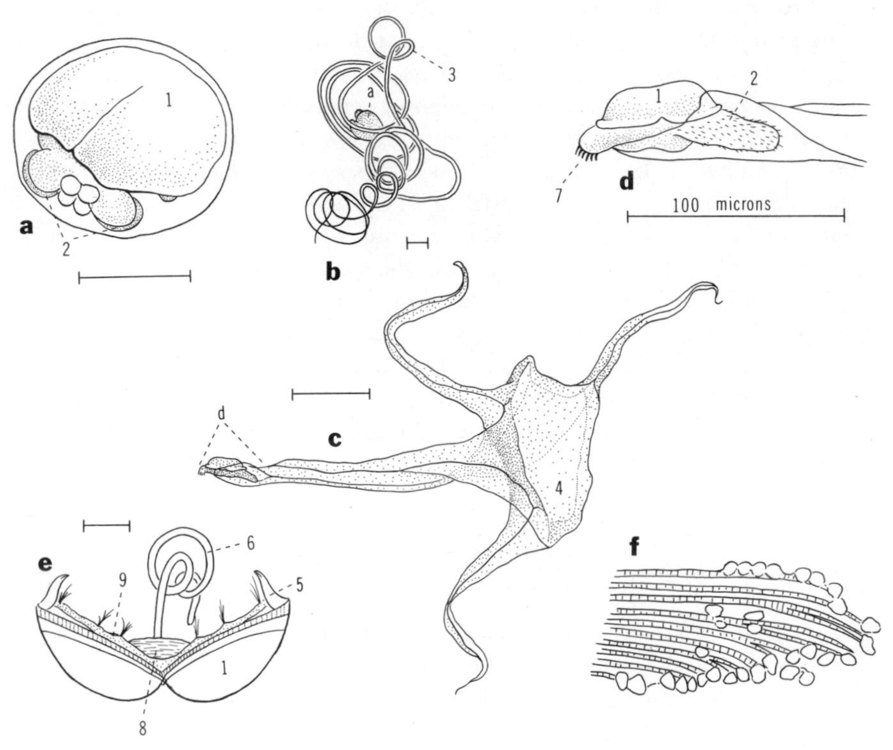

FIGURE 2. Larvae of freshwater unionid clams: (*a*) early haustorium of the African *Mutela bourguignati;* (*b*) later stage of the same species; (*c*) lasidium larva of the South American *Anodontites trapesialis forbesianus;* (*d*) detail of body structure in the same species; (*e*) glochidium larva of the Australian *Velesunio ambiguus;* (*f*) glochidia encysted on the gill of a fish, much enlarged and diagrammatic. Identified structures are larval shell (1), anterior lobe (2), haustoria (3), adhesive organ (4), larval tooth (5), filament (6), hooks (7), adductor muscle (8), and mantle edge (9). *a–b* after Fryer; *c–d* after Bonetto and Excurra; *e* after Parodiz and Bonetto; *f* modified from Lefevre and Curtis.

and closing until they die or come into contact with a fish or salamander. Each clam species has at most a few species that it can parasitize. Frequently only one or two species will be acceptable to the larva. The few lucky larvae clamp onto the gills or body of their host, encyst, and remain as parasites for usually 10 to 20 days, but sometimes as long as six months (Fig. 2*f*). During this time the fish can swim quite a distance. Eventually the larva drops off the gill, sinks to the bottom, and transforms into a juvenile about $\frac{1}{100}$ inch long.

Details differ enormously from species to species, but this basic pattern of a parasitic stage on a fish or salamander characterizes the unionid type clams. It is a wasteful method of reproducing, since only a tiny fraction of the shed "larvae" can hope to encounter a fish body or gill. Once attached, something may eat that fish while the larvae are encysted, or when they fall off they may land on unsuitable bottom. Enormous numbers of larvae must be produced. One study estimated there were 74,000 to 129,000 larvae in a single gravid clam, with a maximum count of 2,225,000 larvae in a specimen of *Leptodea fragilis*. Despite these huge numbers, another study showed that less than 9% of the fish examined were parasitized, although over 30 species of clams were present in that river. There were from 1 to 416 parasites on any single fish, with a maximum of six clam species represented.

Mature females in one group of North American unionids, members of the genus *Lampsilis,* have accessory flaps on the mantle and dark eyespots near their base. Whenever glochidia are ready to be shed, these flaps (Fig. 3) are moved up and down in a way that is characteristic for each species. Undoubtedly this movement would tend to delay settling of the expelled glochidia to the bottom, while to the human eye the appearance of the moving flaps is like that of a small hovering fish. Experimental results were inconclusive, but if a host fish was attracted by these flapping movements, it would be certain to become parasitized by the glochidia.

Since the parasite period of the larva is so important to the dispersal and continued existence of the freshwater unionids, the presence of major differences in the types of larvae has great significance in understanding the evolutionary history of this group. The three larval types are the "haustorium" larva of the Mutelidae, the "lasidium" larva of the Mycetopodidae, and the "glochidium" larva of the Unionacea. The two former types were only discovered and confirmed (respectively) in the 1960s. Far more data are available concerning the glochidium type, known from the mid-1800s, which can be hooked or hookless, with or without the central filament (Fig. 2*e*, 6). The "haustorium" begins as a globular object with anterior ciliated lobes (2) and some posterior hooks (Fig. 2*a,* 7). It fastens to the outside of a

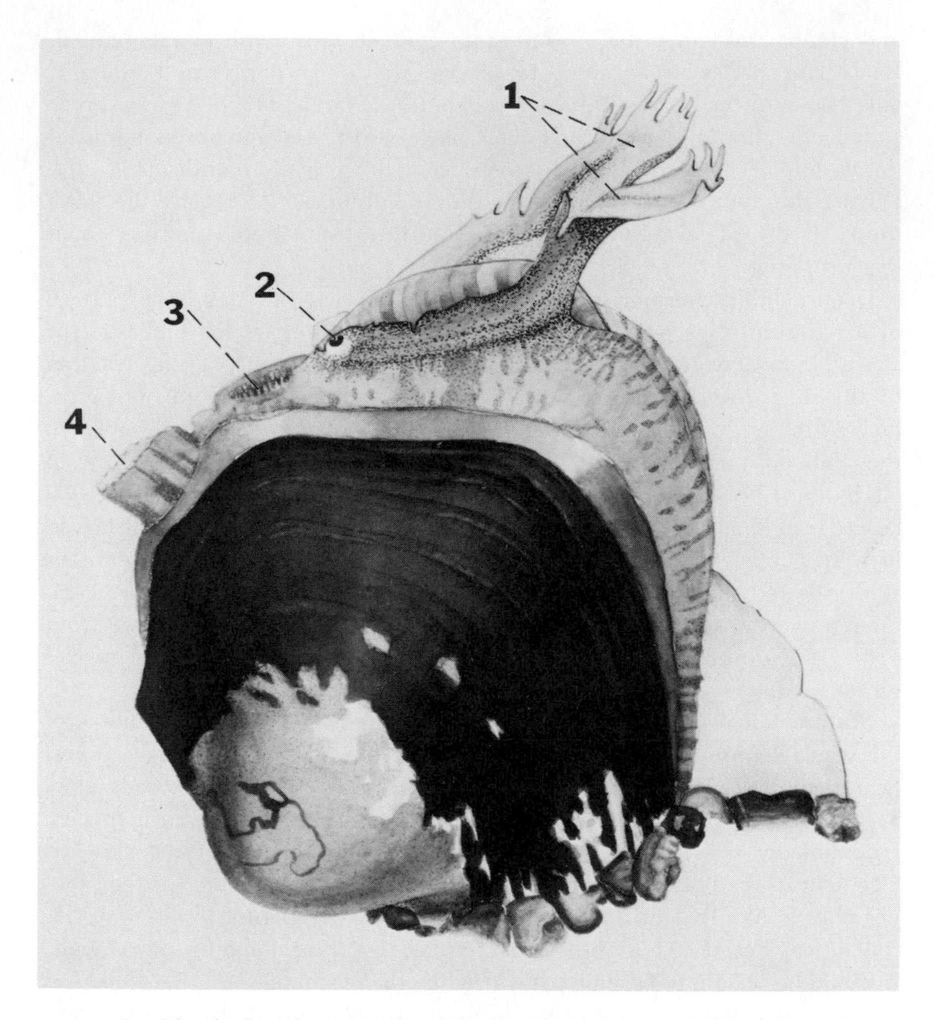

FIGURE 3. Mantle flaps in *Lampsilis siliquoidea* from Arkansas. Identified structures are mantle flaps (1), eyespot (2), inhalent siphon (3) and exhalent siphon (4). Courtesy of Louise Kraemer.

fish, then grows two long tubular "haustoria" (3) that penetrate the tissues of the fish, bind the larva to it, and also serve as a conduit through which fish body juices can be sucked (Fig. 2*b*). The haustorium larva is much larger than the lasidium found in the South American Mycetopodidae. The latter (Fig. 2*c, d*) has a very short body with anterior ciliated lobes (2), caplike shell (1), and posterior hooks (2), but this is attached to an enormous adhesion organ (4) which is used in attaching and encysting. Whether the difference between the glochidium that has two valves when released and the haustorium-lasidium single shell type is significant or merely indicates that the glochidium is retained for a longer period in the female clam is unknown. What is important is that the parasitic larvae of the unionid clams show major differences in structure and mode of functioning.

For unionids to reproduce, both fish and clams must be present. If a clam is put into water without the proper fish, no reproduction is possible. But if a heavily parasitized fish is accidentally carried to a body of water that lacks clams, then a new population of clams can start. Because of their large size and close ties to a fish host, unionids are found mainly in larger bodies of isolated waters or in streams and rivers. Rarely are they found in small ponds. Life history studies of commercially valuable unionids have shown that they may take 7 to 12 years to reach usable size. The total life span is uncertain, although a European species, *Margaritifera margaritifera,* has been aged at 116 years.

As mentioned above, the Eastern United States contained about half of the known unionid species. At the fabled "Mussel Shoals" of the Tennessee River in northern Alabama, before Wilson Dam was constructed, over 70 species of clams lived together. A few headwater streams of the Tennessee-Ohio system still have places where 30 species are found, but the vast majority of shoal dwellers have vanished or are on their way to extinction. Some clam species are adapted to living in the main river channels, and are still abundant, unless killed by pollution. Until World War II they were harvested for cutting into "pearl buttons," but the age of plastics ended this use. In later years the clam fishery was revived. Many Ohio, Tennessee, and Mississippi River species are collected, ground, and tumbled into small pellets, and then shipped to Japan. Inserted into the soft tissue of a living pearl oyster, in a few years they may become the center (nucleus) of a strung cultured pearl that will be exported to the United States.

The best general account of freshwater clams in English remains that given in Baker's *Fresh Water Mollusca of Wisconsin,* but useful identification manuals for several states are listed in Appendix B.

FRESHWATER SNAILS

Many groups of snails have colonized freshwater habitats. The archaeogastropod Family Neritidae, mesogastropods in abundance, and one of the pulmonate superorders, the Basommatophora, are primarily freshwater, but do have some marine and a few semiterrestrial species.

Freshwater snails that have an operculum on their tail are prosobranchs, while those without any operculum are pulmonates. These differ greatly in their strategies for dispersal, distribution, diversity, and economic importance. Most of the prosobranchs remain gill breathers with various devices for ensuring water flow. The peculiar little *Valvata* (Fig. 4*a*) has a featherlike external gill, while the large Ampullariidae (Fig. 5*a*) can use either a gill or breathe air through a long siphonlike tube formed from part of the mantle collar. Since ampullariids frequently inhabit canals or swamps whose waters are very low in oxygen during much of the year, this ability to breathe both ways is very advantageous.

Presence of an operculum also aids survival over drought conditions.

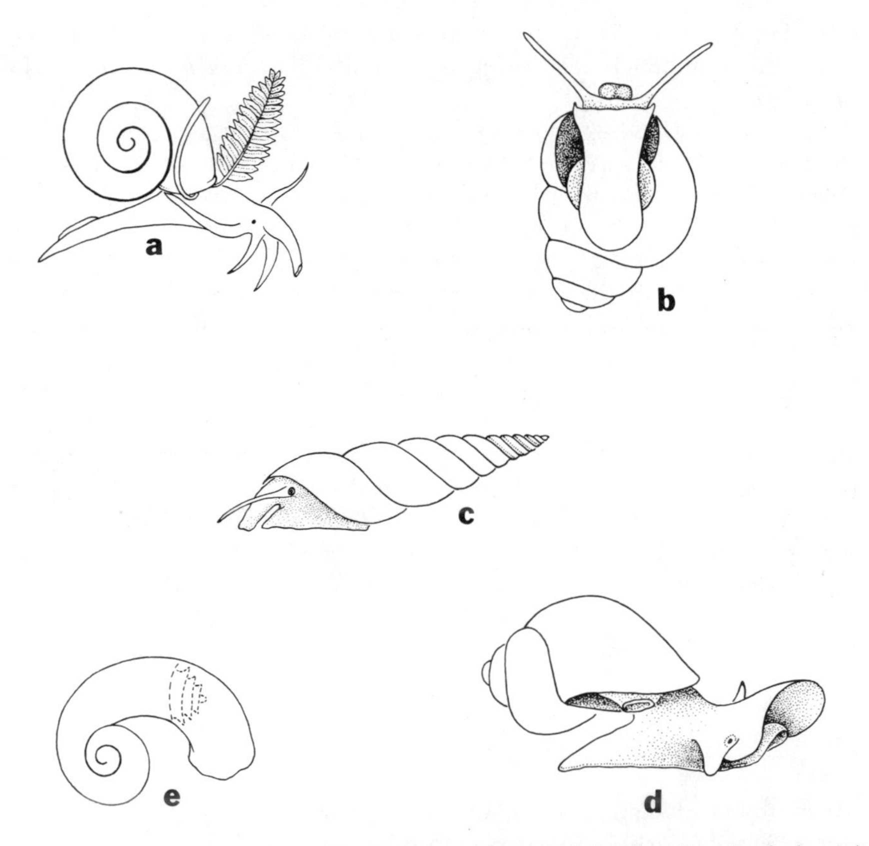

FIGURE 4. Freshwater snails: (*a*) *Valvata* with extended gills; (*b*) *Amnicola;* (*c*) *Pleurocera,* showing its elongated proboscis; (*d*) *Lymnaea,* a typical pulmonate; (*e*) *Gocea ohridiana. a–d* modified from F. C. Baker.

FIGURE 5. Freshwater prosobranchs commonly sold as aquarium snails: (*a*) *Pomacea paludosa* from southern Florida with operculum in place; (*b*) *Cipangopaludina* widely introduced from the Orient into the Northern United States. Shown life size.

Many operculated snails can seal up in dried mud until the monsoon or spring rains return. The effectiveness of this seal can be judged by the experience of a frog collector in Panama. Having an acquisitive malacologist for a friend, he had promised to bring back a few snails. Finding a live *Pomacea cumingii*, a swamp dwelling ampullariid, he plopped it into a solution of alcohol and formalin used for killing frogs. Despite being immersed more than an hour in this solution, the snail later crawled away. Sealed behind its operculum it had not been touched by the deadly liquid.

In most freshwater prosobranchs the operculum is a simple plate, but in a few hydrobiids it is quite elaborate. *Gocea* (Fig. 4*e*) is a minute snail from Lake Ohrid, Yugoslavia, whose shell is "decoiled" for much of its growth. In contrast, the operculum (Fig. 6) is composed of several elevated coils with a fluted edge.

Counterbalancing the advantage of an operculum is the problem caused by prosobranchs having separate sexes. Only a pregnant female snail can populate a new pond or lake if carried to it accidentally. Since most of the prosobranchs are gill breathers, they tend to be most common in running, well-oxygenated waters. Generally they are less common in small isolated bodies of water and still waters except for the swamp dwelling ampullariids.

Pulmonates are hermaphroditic; that is, the individual is both male and female at the same time. In addition they have the ability to fertilize them-

(a)
(b)

selves, so that any snail carried to a new body of water can start a new population. Most species can self-fertilize for two or three generations without difficulty in genetic factors, but some species of *Lymnaea* have had generations of isolated individuals continued until the investigator got tired of the experiment. In every case where it has been tested, however, pulmonate snails will use cross-fertilized sperm in reproduction in preference to sperm from the same individual. They thus can start a new colony with only one snail, but in every possible situation will use cross-fertilization to maintain genetic diversity.

Quite in contrast to the prosobranchs, pulmonates are frequently found in the most isolated and smallest bodies of water. This wide distribution is caused by their tendency to occur on vegetation in very shallow water, air breathing, possession of sticky body mucus, and frequent adhesion to the feet or feathers of water birds. Frequently snails are carried from one pond to another. With their ability to colonize from one specimen and frequent transport, it is no wonder that they are so widely dispersed.

As with the clams and prosobranchs, freshwater pulmonates are most diverse and unusual in the few ancient lakes and older drainage systems. These species, as well as the clams, are subject to extinction caused by pollution. The more adaptable and very widespread species found in temporary ponds and isolated lakes are much less affected by such activities and are in little danger of extinction.

Most freshwater snails are dull brown in color and often heavily encrusted with algae or mineral deposits. A few, such as *Lymnaea stagnalis* (Plate 11, upper left), are somewhat colorful, but most win no beauty prizes and practically none are of direct economic importance.

Indirectly, a very few freshwater snails cause major damage by serving as intermediate hosts for some parasitic worms. These are worms whose eggs are laid in or near water, hatch into larvae that must spend time inside the body of a snail, and then transform into a different type of larva that exits from the snail to seek contact with the skin of its final host. It burrows through the skin and migrates to internal organs, transforms into an adult, and begins to produce eggs, starting a new cycle. Visitors to some Middle Western and European lake resorts may return with a rash known as "swimmers itch." Although rarely serious, it is a bit annoying and of small comfort to the victim to learn that some larval worms that normally infect birds made a mistake. They tried to enter human skin, failed, died, and their

---

FIGURE 6. Operculum of *Gocea ohridiana* (*a*) whole operculum at 200× magnification; (*b*) detail of whorl shoulder at 605×. Courtesy Field Museum of Natural History, Chicago.

decaying corpses caused the rash to appear. More serious are the flukes of cattle and sheep, and most serious are the human parasites or schistosomes. Endemic to much of Africa, the Middle East, Japan, China, and the Philippines, also imported to Brazil and Puerto Rico with the slave trade, "Bilharzia" or "schistosomiasis" probably ranks as the most widespread debilitating disease of humans now that malaria has been partly reduced in frequency. It is inseparable from irrigated agriculture, since the snails that transmit the disease live in paddies or irrigation ditches. Agricultural workers wading in the water will be attacked by larval worms. At the same time adult worms in the human's body will be secreting eggs into the urine or feces (depending on the species). In many peasant societies it is considered "unclean" to urinate or defecate in a toilet, and "proper" to do so in water. So at the same time that humans are releasing eggs into the water to start the parasite's life cycle in a snail, they pick up later larval stages released from the snail. The incidence of this disease in Egyptian villages is over 90%. The snails are essentially impossible to eradicate, and control of this disease must depend on sociological solutions changing the behavior patterns of people, rather than trying to kill off the snails.

In view of the rapid extinction of harmless and helpful freshwater mollusks, it is ironic that the few harmful species have habits which make them nearly impossible to destroy because of their continued accidental transport, ability to self-fertilize, and very rapid reproductive rate.

IX
To Scrape a Living

Previous chapters have focused on the ways in which mollusks locate and capture food. This major preliminary to dinner is followed by getting the food into the body for digestion. This can involve breaking the food into sufficiently small pieces or chunks for swallowing, merely engulfing the whole thing (as with a *Conus*), sucking body fluids from the victim (as do parasites), or using mucus to bind selected small particles into a mass for "swallowing" (as do most filter feeders, clams and snails alike). The last techniques are secondary specializations in mollusks, who use quite a different feeding mechanism than we do.

The upper and lower teeth of other vertebrates and man are used to cut, tear, grind, or macerate food items, and fingers, tableware, and lips are involved in getting food into the mouth. Our teeth are specialized into molars, premolars, canines, and incisors for handling the food, while our tongue and complex muscles move the bits of food about until they can be swallowed. Among the mollusks, only the cephalopods use a roughly similar system in that tentacles (= arms) grasp and pull food to the mouth, then a pair of strong beaks or jaws (= our teeth) tear the food into pieces that enter the mouth. Scaphopods have tentaclelike feeding arms, but do not have a jaw apparatus. Excluding the filter feeding clams and snails,

FIGURE 1. Patterns of tooth wear in pulmonate land snails; (*a*) unworn lateral teeth of *Suteria ide* from North Island, New Zealand, at 2400×; (*b*) teeth from anterior end of the same radula that have been chipped down to uselessness, at 2400×; (*c*) unworn lateral teeth of the New Ireland, Bismarck Archipelago species *Papuina phaeostoma,* at 1100×; (*d*) badly eroded teeth from the anterior of the same radula, at 1250×; (*e*) unworn marginal teeth from *Succinea norfolkensis* collected on Mt. Pitt, Norfolk Island, South Pacific, at 5000×; (*f*) surface of worn lateral tooth on the same radula, at 2425×. Courtesy Field Museum of Natural History, Chicago.

(a)
(b)

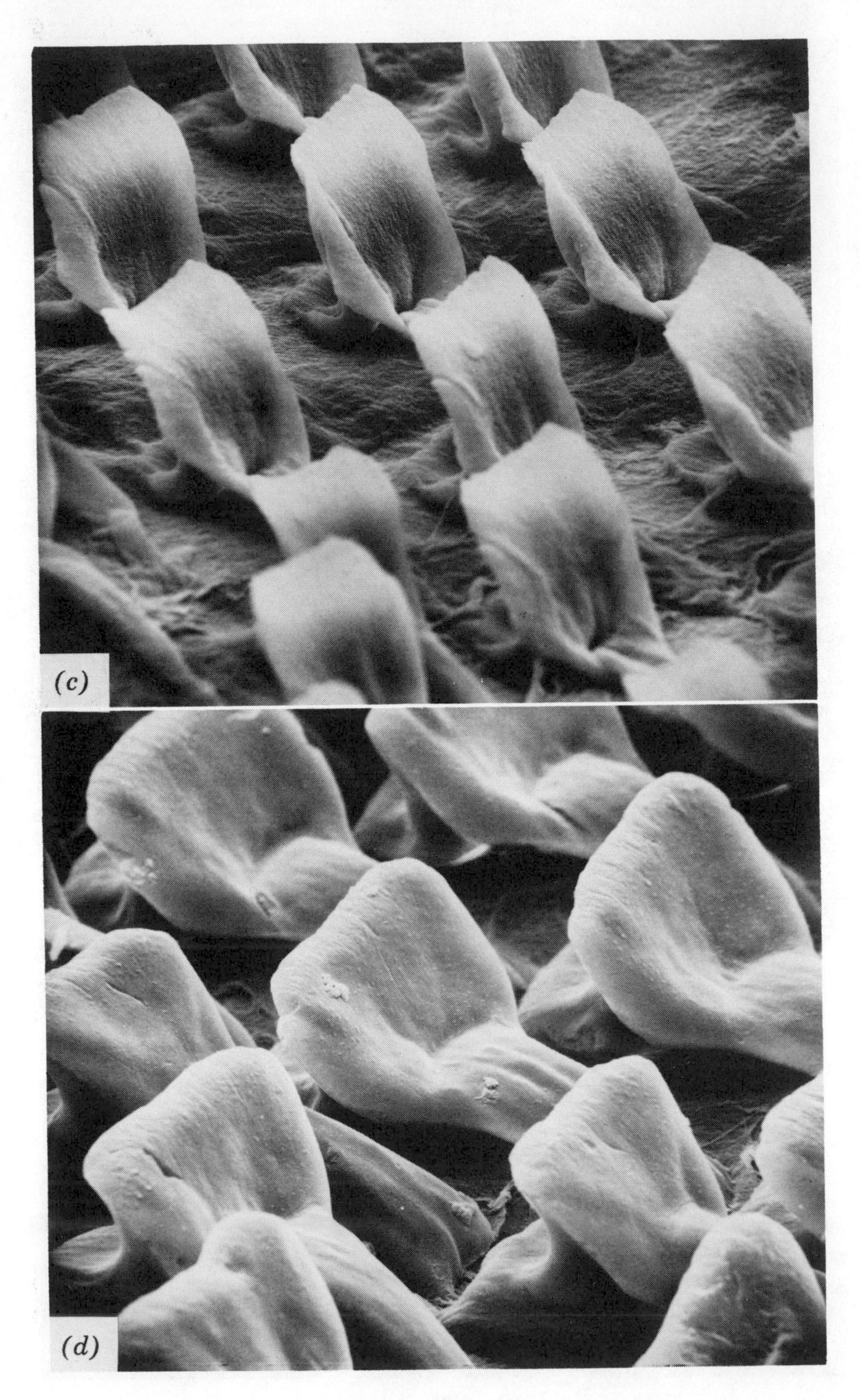
(c)
(d)

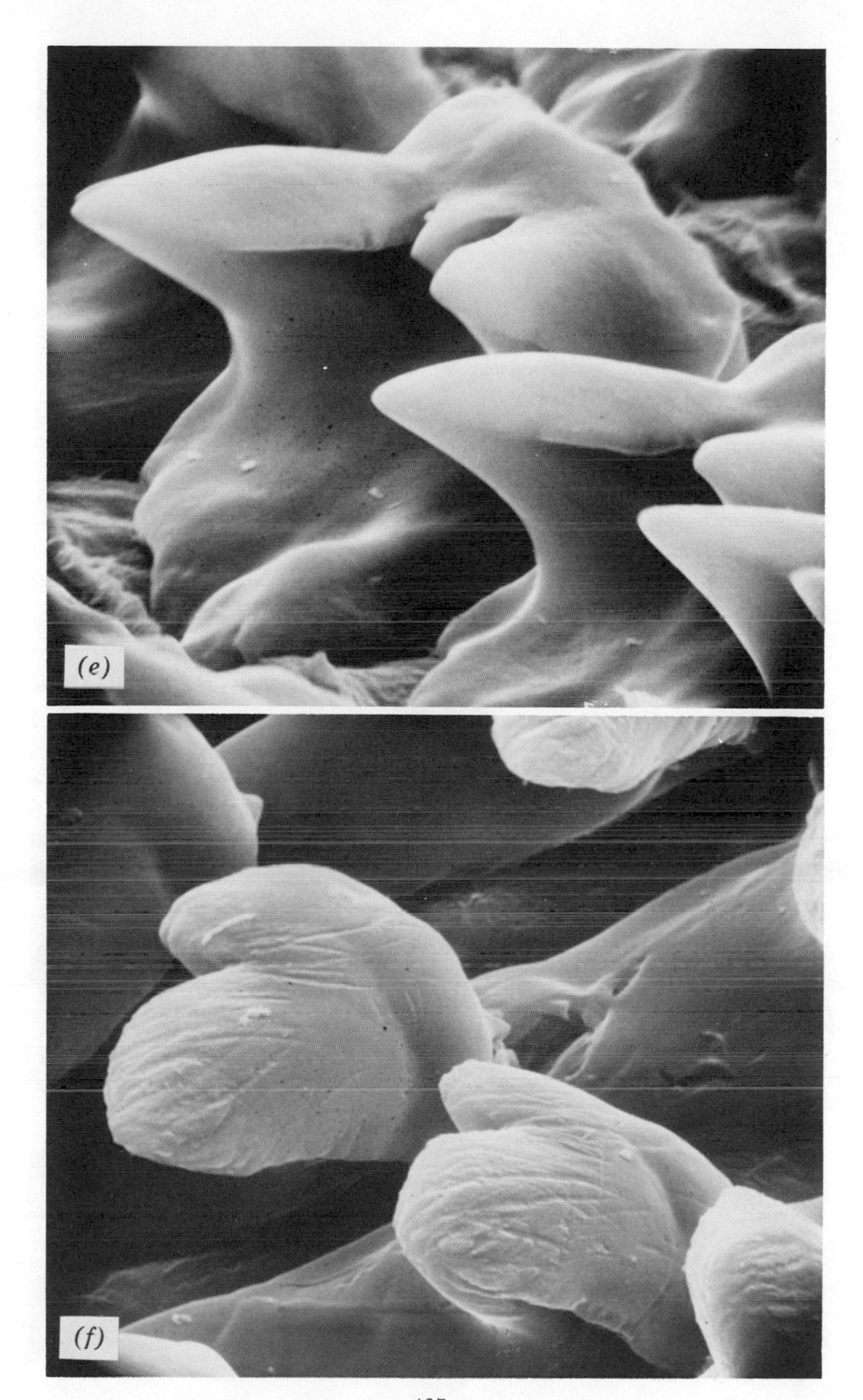
(e)
(f)

(a)

c
cc
r
a
(b)

FIGURE 2. Radular teeth of a land archaeogastropod, *Sturanyella plicatilis,* from Upolu Island, Western Samoa: (*a*) rows of marginal teeth, at 445×; (*b*) part of central and lateral teeth complex, at 520×; (*c*) details of lateral teeth, at 685×. Identified structures are R-central (r), A-central (a), B-central (b), C-central (c), and capituliform complex (cc). Courtesy Field Museum of Natural History, Chicago.

all generalized mollusks depend on protruding a tonguelike organ coated with a few to many thousands of teeth on its surface and rasping or tearing loose pieces of food. This protrusible organ or *radula* both tears or scrapes *and* pulls food particles into the mouth. In snails it may or may not be assisted by one or more jaws. This is quite a different method of feeding than that used by the vertebrates, and has both advantages and limitations.

A radula is absent in clams and highly modified in parasites and carnivores such as toxoglossans. In its most general form, the radula consists of a membrane on which are mounted rows of teeth that point backward. Complex muscles control in and out movement of the radula and associated supporting cartilages, during which the radular membrane is rotated partly over

FIGURE 3. Radular teeth of a hydrobiid freshwater snail from Lake Ohrid, Yugoslavia: (*a*) low angle view of a partial tooth row, at 4100×; (*b*) central tooth, at 14750×; (*c*) several lateral teeth, at 5350×; (*d*) cusps on the inner (large) and outer (small) marginal teeth, at 7275×. Courtesy Field Museum of Natural History, Chicago.

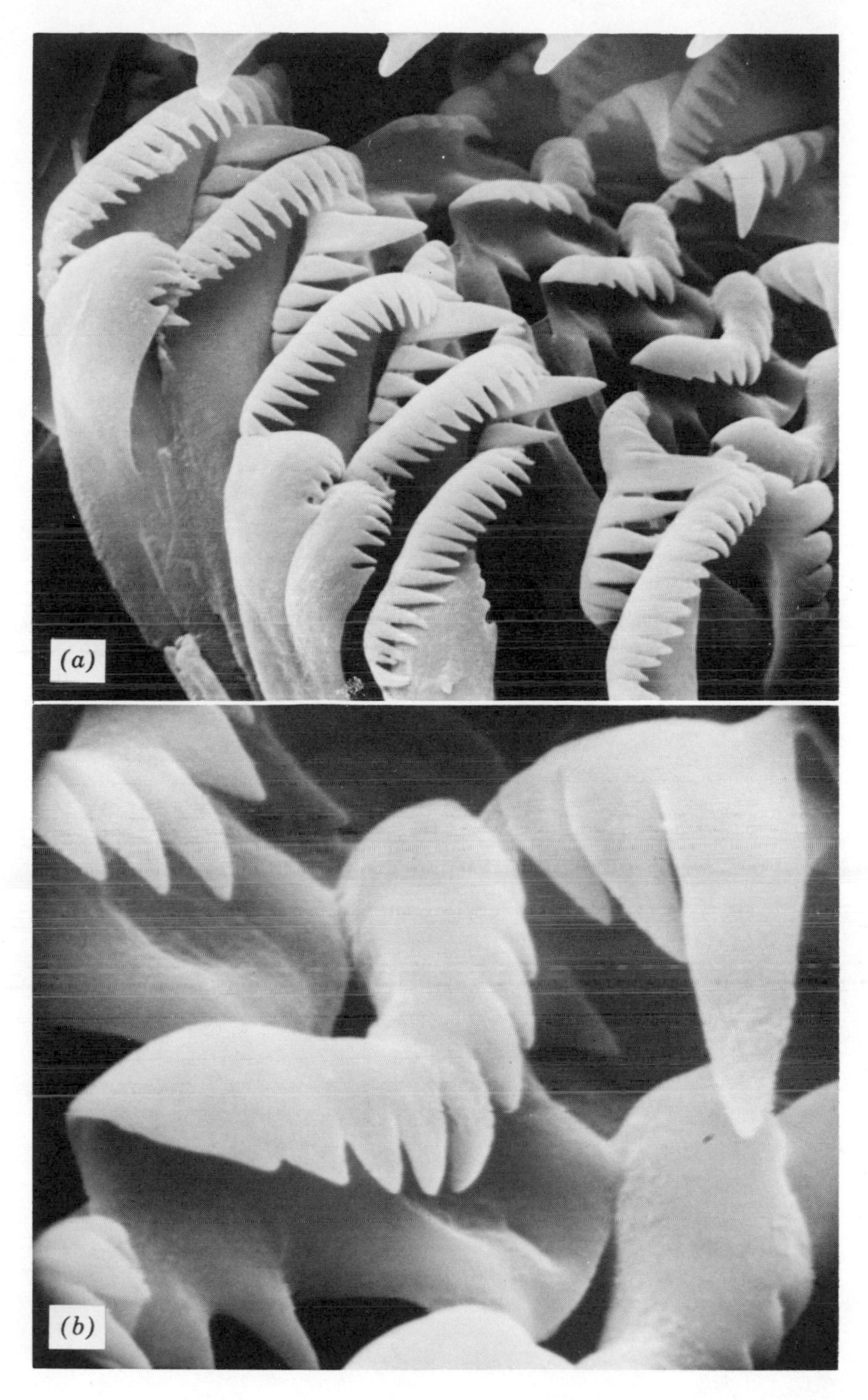
(a)
(b)

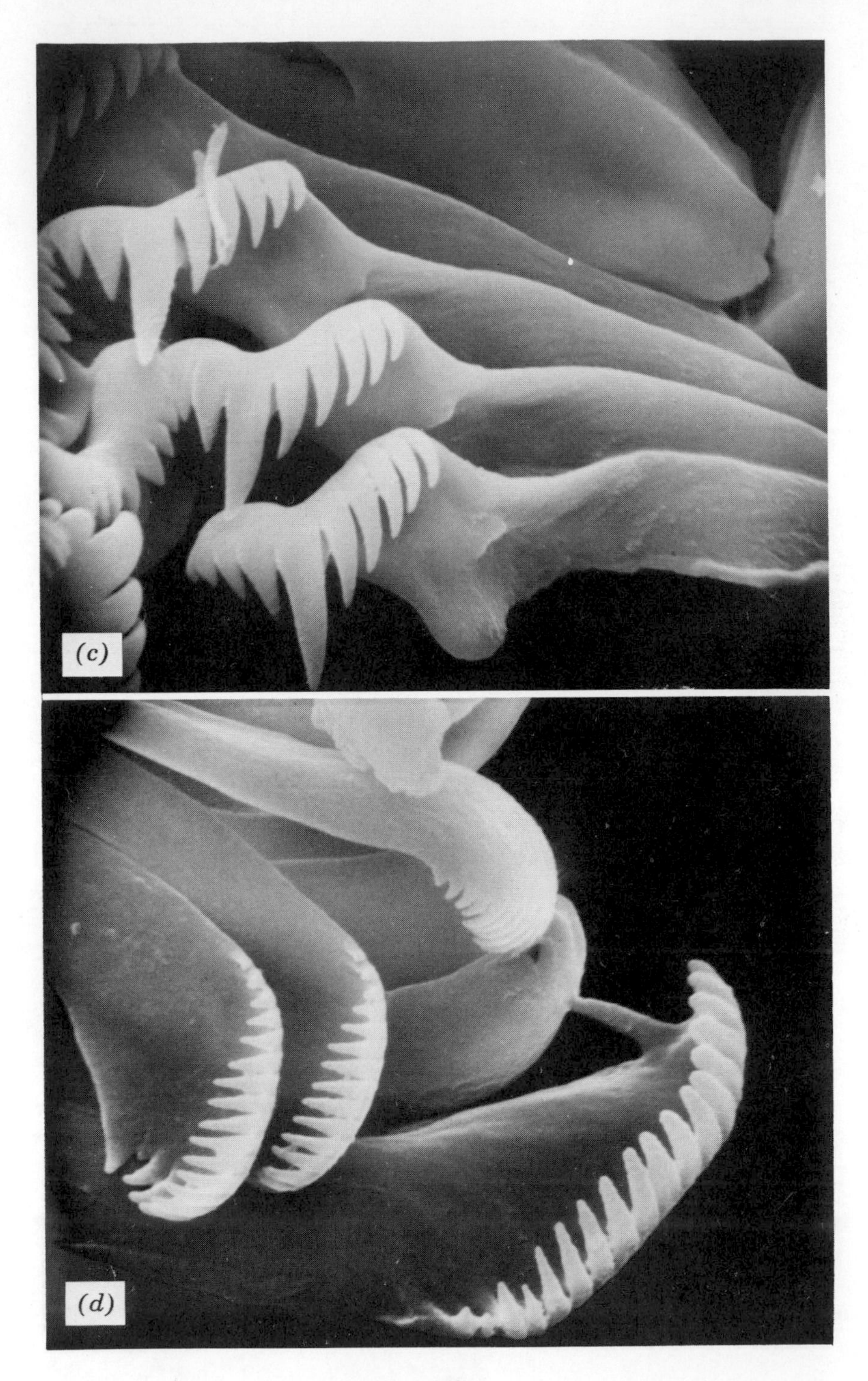
(c)
(d)

the tips of the underlying support. When protruded the teeth on the radula can scrape, pierce, cut, or tear the object with which they come in contact and produce small bits for swallowing. Often some teeth will be specialized for breaking up the food, while others catch and pull the pieces into the mouth.

A major difficulty is that teeth protruded in a random licking or scraping motion will become gouged, chipped, or ground down either by the food itself or by a hard surface from which food is scraped. In simplest form, each tooth consists of a *basal plate,* by which it is attached to the radular membrane, and one or more elevated sharp edges or *cusps.* Figure 1 shows patterns of wear in pulmonate land snail (pp. 135–137). The top illustration of each pair shows unworn teeth, the bottom figure worn teeth from a different part of the same radula. In the New Zealand endodontid *Suteria ide* (Fig. 1*a, b*) the teeth become chipped and broken off; in the Bismarck Archipelago tree snail *Papuina* (Fig. 1*c, d*) the single broad cusp on each tooth wears down to a nub; while in a Norfolk Island species of *Succinea* (Fig. 1*e, f*) surface grooves on worn teeth tell of gradual abrasion to their soft surface. This wear problem is solved by having new rows of teeth formed at the posterior end of the radula throughout the life of the animal, while worn teeth from the anterior end are continuously falling off to be swallowed and passed out in the feces. The whole radular ribbon with its rows of teeth grows forward at a rate of five to six transverse rows per day for browsing species such as *Littorina* or plant eating slugs, but only one to two rows per day in some carnivores. The exact rate of replacement growth varies with the species and environmental factors. The fact of this continuous loss and replacement of worn teeth explains one great advantage of the molluscan radula.

In the aplacophorans, radular teeth are not mounted on a membrane, but are placed directly on the walls of the mouth in various arrangements. Chitons have a more typical radula with 17 teeth in each row, but there is no jaw. Scaphopods have a simple radula with a few teeth that are used to tear up food captured by their tentacles. Cephalopods have a pair of beaks or jaws that are used to bite or tear their food, with the radular teeth assuming less importance. Monoplacophorans have 11 teeth in each row, two of which have long comblike denticles extending horizontally from the upper edge. It is in the gastropods that radular diversity is the greatest.

Literally thousands of drawings have been made of gastropod radular cusps over the years, but their complex form and usually small size have meant that understanding their structure and function was extremely difficult. The availability in the late 1960s of a new research tool, the scanning electron microscope, has permitted considerable advances in knowledge and

FIGURE 4. Mesogastropod radular teeth in species of the genus *Ostodes* found on Upolu Island, Western Samoa: (*a*) a row of teeth in relaxed but partly opened position, at 500×; (*b*) central and right lateral tooth showing how it can be folded in next to the central tooth, at 1150×; (*c*) left marginal teeth, at 1075× showing cusp structure; (*d*) lateral and marginal teeth, at 555×, elevated for a feeding stroke. Identified structures are central (c), lateral (l) and marginal (m) teeth. Courtesy Field Museum of Natural History, Chicago.

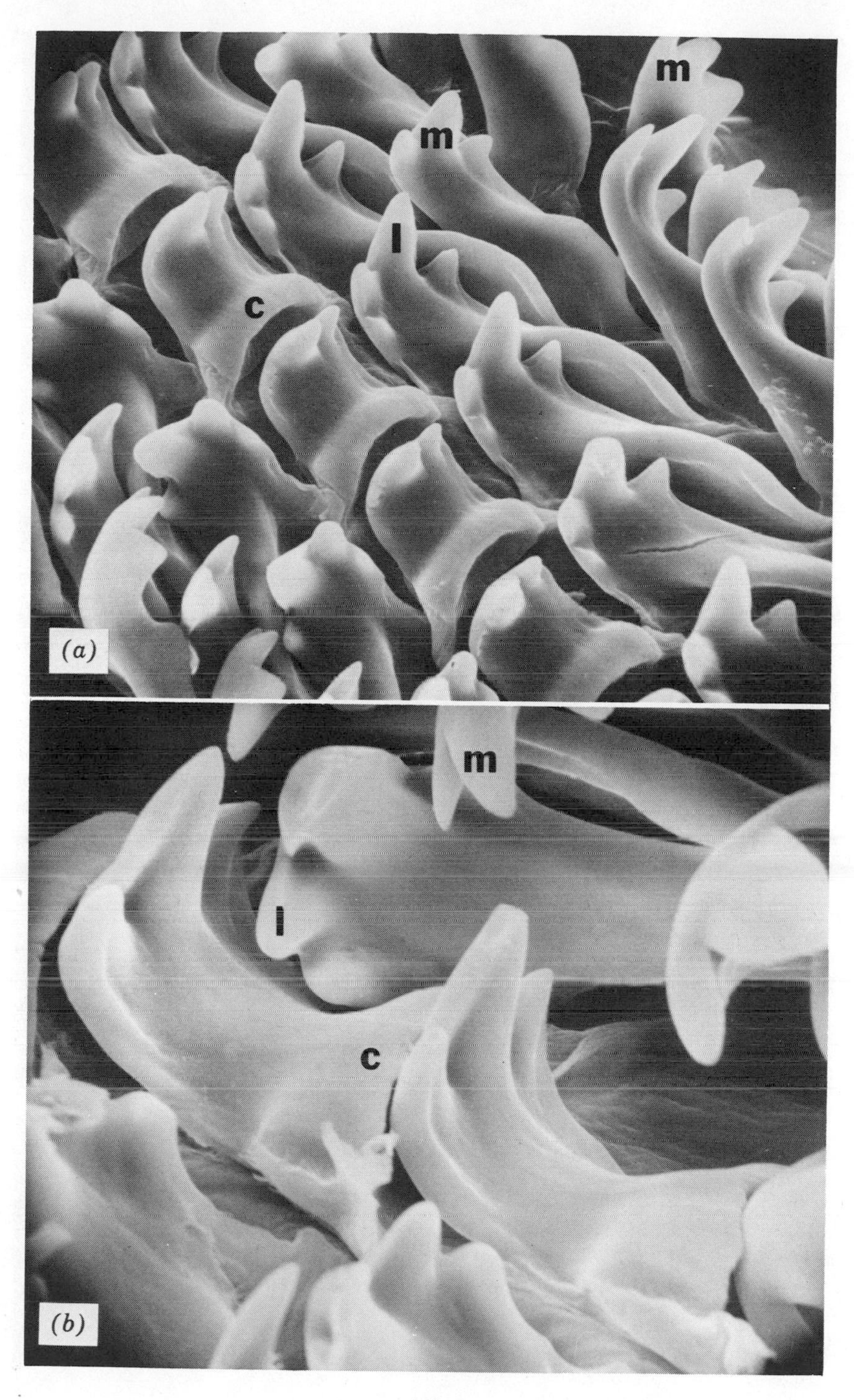
m
m
l
c
(a)
m
l
c
(b)

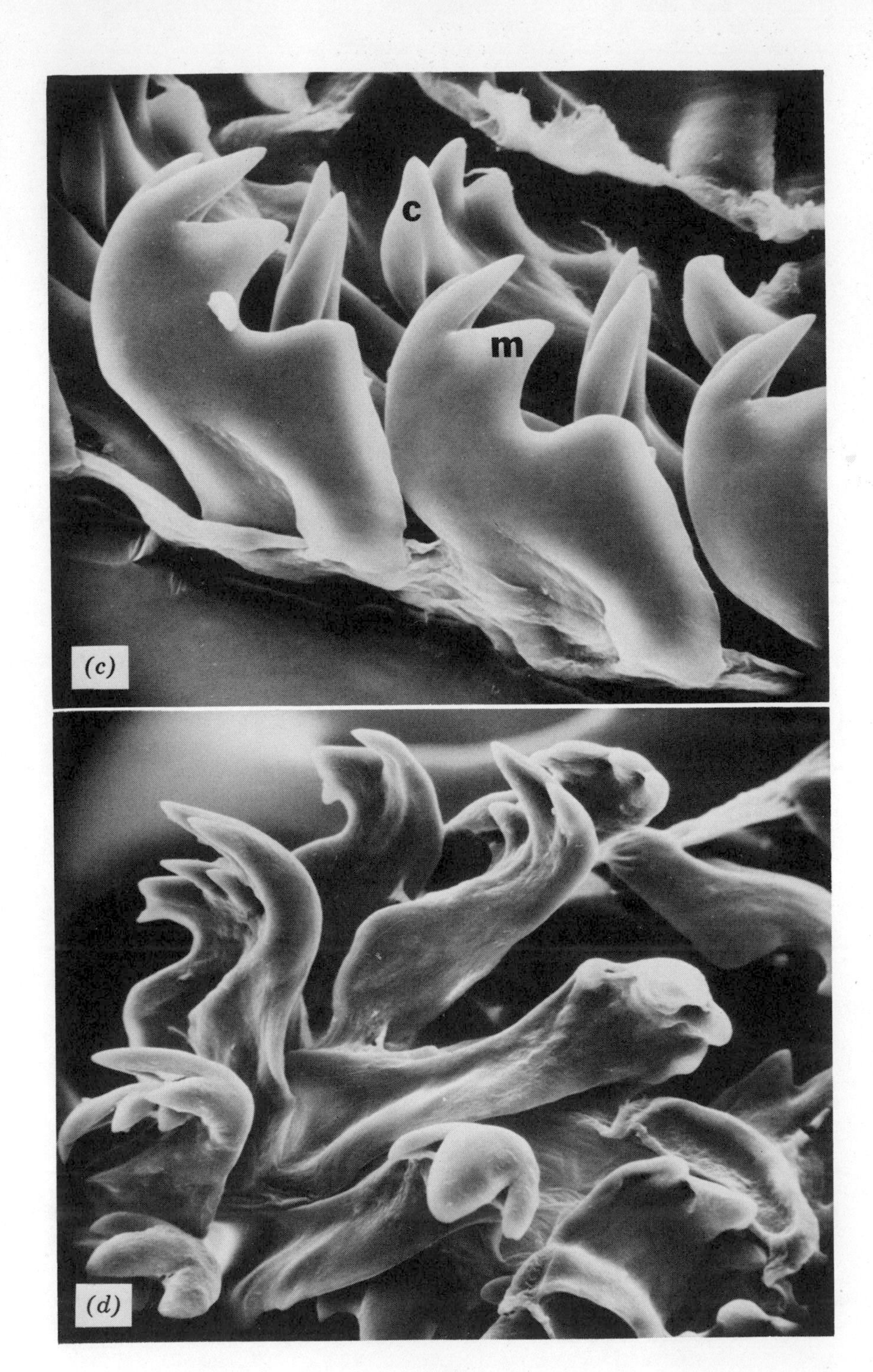
c
m
(c)
(d)

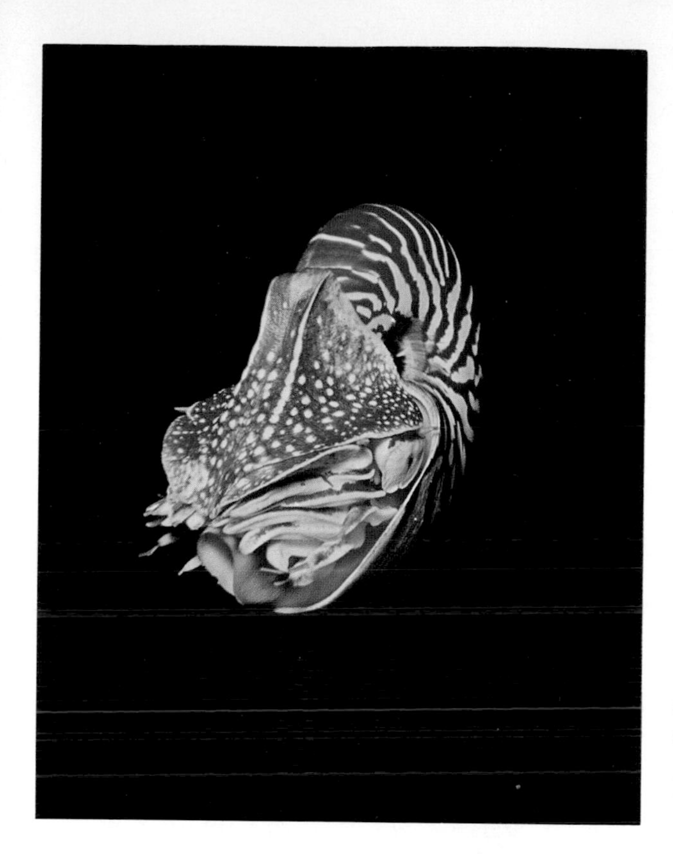

PLATE 1. Chambered nautilus. One of the half dozen living species of nautiloid cephalopods, *Nautilus macromphalus* is found at moderate depths along the reefs off New Caledonia in the South Pacific. The upper photograph shows a swimming specimen with many of the *cirri* mostly retracted into the tentacles. The hyponome or funnel is twisted toward the left of the picture to jet the animal away at an angle. At the right margin, where the hood and shell meet, the large eye with vertical slit and two small ocular tentacles are clearly visible. In the lower photograph, taken at a depth of 70 feet while diving at night, tentacular cirri are extended. Two of them are gripping the rock surface, while the lower tentacles are spread in a "cone of search" ready to grab any potential food items. (Photographs by Douglas Faulkner.)

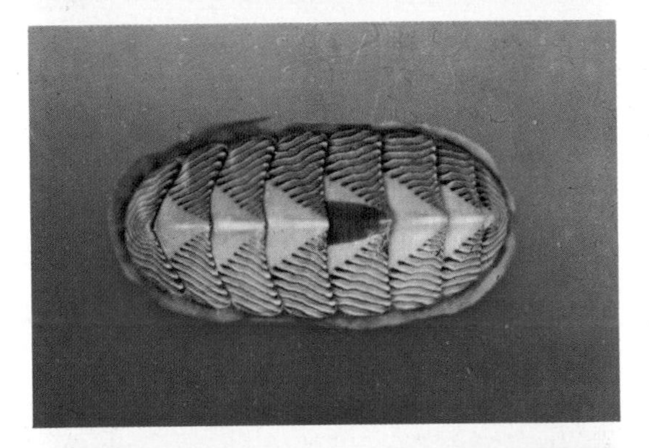

PLATE 2. Clams and chiton shells. *Upper left: Pitar dione,* a Caribbean burrowing suspension feeder. The long shell spines on each valve are thought to help protect the clam's siphons from being bitten off by a fish. Length of shell including spines is about 2 inches. *Upper right: Corculum cardissa,* an Indo-Pacific clam whose shell is greatly compressed anteriorly and posteriorly, creating a shell much thicker than it is long. The shell sculpture, contours, and color are quite different at the anterior (left) and posterior (right) ends. Length of shell is slightly less than 2 inches. *Center right: Tonicella lineata,* a 2 inch long chiton from Western North America. The colors are somewhat faded from those seen in living specimens. *Lower right: Spondylus americanus,* a 6 inch specimen from reefs off the Florida Keys. The lower valve is anchored to the reef surface. At the extreme right, the flat white portion is the "cardinal area," where the center groove represents what is left of an external ligament. The internal ligament is shown in Fig. 4*d* of Chapter V.

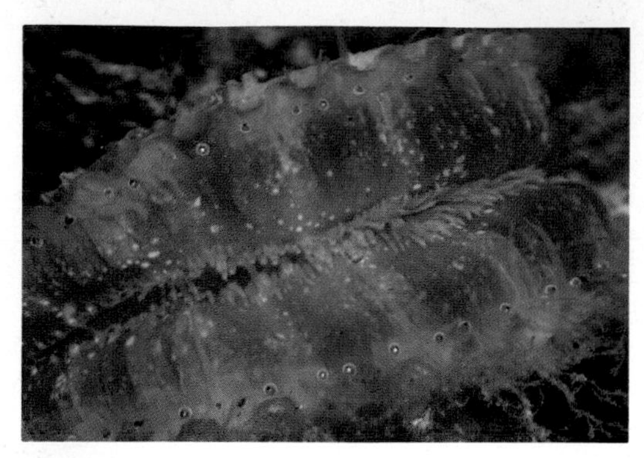

PLATE 3. Shells and shell makers. *Upper right:* Mantle edge of the pectinid *Chlamys gloriosa.* This Great Barrier Reef species has bright blue eyes rimming its mantle. Numerous small lobes lines the inner mantle edges to prevent silt or sand from fouling its gills. *Upper left: Chlamys gloriosa,* from Keppel Bay, Queensland, Australia. The shells have gaped open only about half as wide as they can. If the shell valves are snapped shut suddenly, the clam will "swim" by jet propulsion. Shell color variation is normal in this group. *Center left:* An overturned Queensland *Lambis lambis,* about 8 inches long, has begun to emerge from its shell. The yellow-spotted tentacles and tubular proboscis are seen at far right. The spadeshaped anterior propodium of the foot is next, followed by the metapodium, which ends in the claw-like, brownish operculum. The latter will be dug into the bottom and used to lever the shell back into a more normal position. *Lower left:* Specimens of *Lambis crocata* from the Philippines and a 2 inch diameter *Astraea phoebia* (right) from the Florida Keys present a contrast in spine structure and position. The *Lambis* develops such spines only when adult; the *Astraea* has the edge of its shell ridged with these spines throughout its life. (Photographs of living *Chlamys* and *Lambis* by Don Byrne.)

PLATE 4. Marine snail types and a used-shell user. Prosobranch snails are divided into three great groups—the primitive archaeogastropods; mesogastropods, with a browsing or carnivorous habit; and the stenoglossan scavengers, borers, or poison equipped predators. Members of each group are shown here, together with one of the inheritors of used snail shells, a hermit crab. *Top: Haliotis asinina,* a small (4 inches long) abalone found on the Great Barrier Reef of Australia at Masthead Island, Central Queensland. The leaf-like flaps are expansions of the epipodial fringe seen in many archaeogastropods (see Fig. 3*a* in Chapter VI). An outline drawing of the shell is shown in Fig. 3*c* in Chapter VI. *Second from top: Ficus subintermedius,* a carnivorous mesogastropod from the Great Barrier Reef that feeds on echinoderms. The 3-inch shell is elongated into a protective canal for the siphon and extends in front of the head. *Third from top:* The White-spotted Hermit Crab, *Dardanus megistos,* looking out from its home in the shell of a dead *Turbo.* Dead shells can take years to disintegrate into small fragments. In the meantime, many generations of hermit crabs make use of spiral snail shells to protect their vulnerable abdomen and tail. Only the heavily armored head and claws are exposed to possible attack. *Bottom:* A toxoglossan hunter armed with spear and poison gland, *Conus ammiralis* has a 2½ to 3 inch shell whose color pattern is clearly visible. The orange-tipped siphon (far right) is used to sample inflowing water for traces of chemicals from a possible "meal" in addition to pumping respiration currents. (Photographs by Don Byrne.)

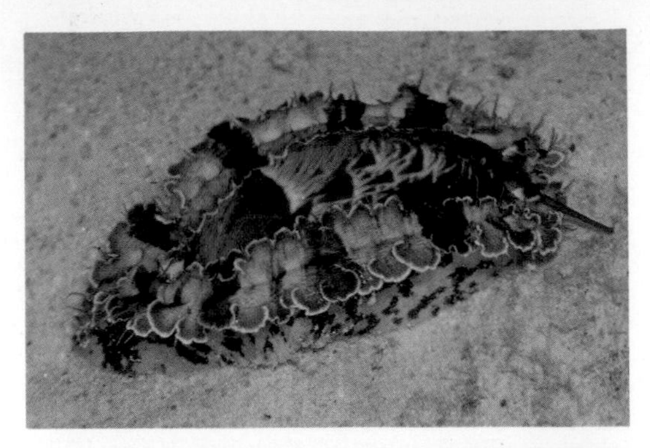

PLATE 5. Snail shells and their makers. A shell can give protection in many ways. Formation of sharp spines to discourage an enemy, building such a massive shell that nothing can break it open, or having the shell surface overgrown with tiny organisms to camouflage it from enemies—all these are effective means. Here are examples of each technique. *Top:* Living *Murex acanthostephes* from Queensland has a short stocky foot and slim tail (left of foot stalk) on which sits the horny operculum. The siphon extends through the long shell tube to the right, letting the *Murex* investigate areas far in front of its own head. (Photograph by Don Byrne.) *Second from top:* Shell of a 6 inch *Murex pecten* with the operculum sealing the aperture. This view gives a better idea of the length and variety of shell spines. *Third from top and bottom:* Two shells of the intertidal muricid *Drupina grossularia* blend into the rock surface (third from top) except for their shadows. When turned over, their orange shell and white lip nodes provide a striking contrast. The white "bumps" in the background are many small barnacles. These 1½ inch shells were photographed on rocks in the tidal zone at Upolu, Western Samoa.

PLATE 6. Living cowries of Queensland, Australia. Cowry shells have been used as symbols of chieftainship, strung as money, and made into decorative art in numerous societies. They have been favorites of shell collectors for 300 years. The animals are even more colorful than the shells, as the species shown here demonstrate. They are $1\frac{1}{4}$ to $1\frac{1}{2}$ inches in length. *Upper left: Cypraea saulae nugata* has only a few scattered yellow dendritic (branched) papillae on an orange mantle. The right-hand specimen has the mantle less extended. *Upper right: Calpurnus verrucosus* is an egg cowry, not a true cowry. It has a pure white shell, but the mantle (crowded black spots) and foot (scattered black dots) are highly distinctive. *Center and lower right: Cypraea cribraria* with expanded (center) and partly contracted (lower) mantle presents a brilliant contrast between shell and animal. This probably serves to confuse a predator. If a touched mantle is withdrawn, the shell is very different in color and texture. Note the short red siphon protruding from the shell. (Photographs by Don Byrne.)

PLATE 7. Cowries and volutes from Queensland, Australia. *Upper right: Cypraea subviridis* has a dark brown body and mantle, which is withdrawn in this view. The whitish mass on the left consists of egg capsules laid by the cowry. The rock has been turned over so they can be photographed. *Upper left: Cypraea limacina facifer* has white spots on a light to dark purplish shell. The mantle is orange with crowded, quite long papillae. The siphon (far right) protruding from the shell is far longer than the siphon of *Cypraea cribraria* (Plate 6, center and lower right). *Center left:* There is little contrast between the shell and animal of *Cymbiolacca pulchra,* a 3½ inch species from coastal Queensland. It is not known if the small spines on the shell have any function. *Lower left:* A juvenile *Melo amphora* sits on a rock. The very large white nuclear whorls (right) indicate that this species hatches as a crawling young. (Photographs by Don Byrne.)

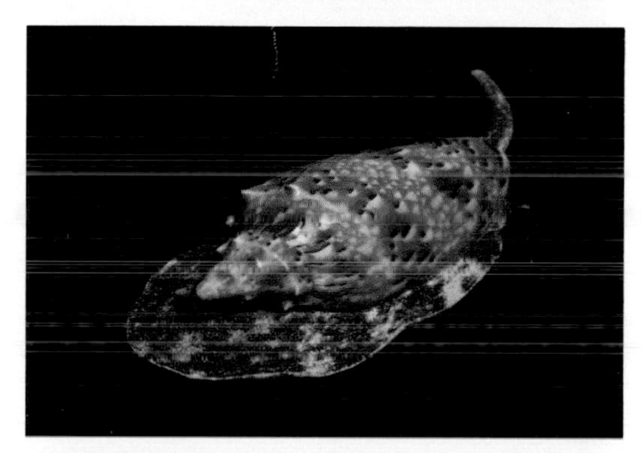

PLATE 8. Living volutes of Australia. Volutes are shallow to deep water snails which display a variety of color in both animal and shell that is unrivaled by any group. Many can be collected only by dredging or trawling, and have very limited distributions. Volutes are among the most expensive of all shells treasured by collectors. *Top and second from top: Amoria maculata* is a $3\frac{1}{2}$ to 4 inch shell found in shallow waters off Queensland, Australia. The extended siphon and partly protruded proboscis are visible in front view (top). The black spot near the base of the tentacle is the eye. Side view of the animal in an aquarium (second from top) shows its crawling posture. The siphon is twisted up toward the photographer. *Third from top and bottom: Amoria canaliculata* is a $2\frac{1}{4}$ inch deep water species trawled off Queensland. The proboscis (bottom) is more fully extended and the coloration of the animal far more dramatic and very different from that of *A. maculata,* despite their belonging to the same genus. The name *canaliculata* refers to the sutural groove on the shell spire shown in the photograph third from top. (Photographs by Don Byrne.)

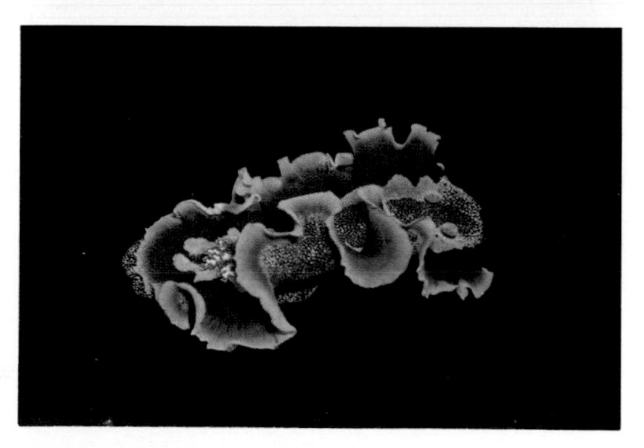

PLATE 9. Sea slugs. Ranging from cryptically colored specimens impossible to separate from the algae on which they crawl unless they move, to brilliant warning colorations, sea slugs, members of the Order Nudibranchia, are little studied. The Queensland species shown here are of uncertain identification, but demonstrate clearly the varied coloration and beauty of this group. *Top: Phyllidia varicosa* has orange tentacles and white or orange "warts" on its back. The irregular color markings are typical of this group. *Second from top:* A colorful striped dorid nudibranch sits near its strip of eggs. The latter may contain literally millions of eggs that hatch into swimming larvae. *Third from top:* Similar looking egg strings are laid by this bulky white sea slug. The erect projections at the right are the cerata, used by the nudibranch in respiration. In the dorid just above, the cerata and tentacles are brownish. *Bottom: Hexabranchus* can reach almost 10 inches in length. It swims by undulating the red and white fringed parapodia on each side of its body. At the left is the cluster of cerata and on the right the two buttonlike tentacles. (Photographs by Don Byrne.)

PLATE 10. Land snails and sluglike creatures. Snails have moved from water to land more than 10 times, producing a variety of color and structure only rarely glimpsed by collectors or biologists. A typical land prosobranch, an advanced pulmonate, and a sluglike species are shown here. *Upper left: Ostodes tiara,* a 1 inch mesogastropod found in the mountains of Upolu, Western Samoa. The operculum (brown disk) and eye spots (at base of red tentacles) are clearly visible. *Upper right: Meridolum gulosum,* a camaenid pulmonate land snail from near Sydney, Australia. Its shell is about $1\frac{1}{2}$ inches in diameter. *Center right:* A semislug belonging to the genus *Helicarion* from Yeppoon, Queensland, Australia. The mantle lobes are partly contracted and the animal obviously is too big to retreat into its small shell. *Lower right:* A *Helicarion* with the mantle lobes expanded to cover the shell almost completely. Total body length would be about $2\frac{1}{2}$ or 3 inches. (Photographs of *Helicarion* by Don Byrne.)

PLATE 11. Form and color in nonmarine shells. The flared lip of a brown freshwater snail contrasts with the brilliant colors of tree snails. *Upper left: Lymnaea stagnalis,* a freshwater snail from the Northern United States, Canada, and Europe. *Upper right: Liguus fasciatus solidus,* form *pictus.* A 3 inch tree snail from Big Pine Key, Florida Keys that has been transplanted into Everglades National Park to ensure its survival. This is one of the rarest and most sought-after color forms of *Liguus. Lower left:* Tree snails of the genus *Polymita,* from Oriente Province, Cuba show an unrivaled variety of banding and tones.

PLATE 12. Pacific island land and tree snails. *Top:* A New Zealand carnivore, *Paryphanta busbyi,* feeds on earthworms and slugs. The black body stretches far out from its 3 inch diameter shell. *Second from top:* A $\frac{1}{2}$ inch long *Achatinella juddii,* a tree snail from middle altitudes on the Koolau Mountains, Oahu, Hawaiian Islands. The short head and tail are characteristic of plant or algae eating pulmonate land snails. *Third from top:* Three specimens of a *Trochomorpha* from Fagaloa Bay, Upolu, Western Samoa. This helicarionid land pulmonate lives on the moss covered trunks of trees and is about 1 inch in diameter. *Bottom:* Color variation in *Achatinella mustelina* from the Waianae Mountains of Oahu, Hawaii. The lihua bush is a favorite habitat; the snails seal to the underside of a leaf during daylight hours.

greatly simplified illustration problems. Photographs made in research on land snail feeding patterns are used to demonstrate this diversity and some of the ways in which the teeth function. An important fact to grasp, particularly in regard to large or complex teeth, is that snail teeth function when they are protruded from the mouth and are being rotated around the tip of the supporting cartilages. In this position they are elevated and used to cut, slice, or pull. The rest of the time they must be folded down or otherwise compacted so that they do not injure the inside of the snail's mouth. This is particularly a problem for carnivores (Fig. 10) and those prosobranchs with high and complex teeth (Figs. 2 to 4). A second important fact is that the teeth interact with each other during functioning—they are not independent.

Helicinid land snails have a *rhipidoglossate* radula (Fig. 2, p. 138), known by huge numbers of small *marginal* teeth (Fig. 2*a*). These are used to catch and pull food particles into the mouth after they have been torn or ripped loose by the much larger *lateral* teeth (Fig. 2*b, c*). In the middle of the radula is a single *R-central* tooth flanked on each side by four mirror-image teeth before the marginals begin. The radula is symmetrical, having marginal-lateral-central-lateral-marginal teeth from left to right in each *transverse* row. The teeth also line up in *longitudinal* rows, all teeth being the same. The Samoan species whose radula is shown, *Sturanyella plicatilis,* is common on tree trunks and on the leaves of banana plants where it feeds by scraping off the algal film growing under these wet tropical conditions. The middle tooth has an unserrated upper edge, and, when the radula is curved as in a feeding stroke (Fig. 2*b*), it seems to act as a support for the teeth on either side. It is called an "R-central" tooth (labeled r). On either side is the three- or four-cusped "A-central" tooth (labeled a), followed by a six- or seven-cusped "B-central" tooth (Fig. 2*c*, b). A smaller, spoon-shaped tooth called the "C-central" (c) has the cusps on its upper edge greatly reduced in size and serves to support the edge of the huge and multicuspid "capituliform complex" (cc). This complex apparently does most of the actual feeding work. As can be seen from the torn edge view (Fig. 2*c*), the shape and interrelationships of these teeth are very complicated. Except for the limpets, which have a greatly modified type of tooth, this rhipidoglossate radula is characteristic of archaeogastropods.

Most mesogastropods have a taenioglossate radula, with only seven teeth in each row—a central flanked on either side by one lateral and two marginal teeth. Within this number restriction, great diversity of form is possible. The minute hydrobiid, *Gocea ohridiana* (Fig. 4*e* in Chapter VIII), from Lake Ohrid in Yugoslavia has a complex radular structure. Seen from a low angle (Fig. 3*a*), the radula is a confusing mass of cusps. If isolated

FIGURE 5. Rachiglossate radula of the stenoglossan prosobranch *Drupa ricinus* from Sigatoka, Viti Levu Island, Fiji, collected on intertidal rocks: (*a*) looking anteriorly at central teeth along a curved portion of the radula, at 310×; (*b*) central and lateral teeth, at 420×. Courtesy Field Museum of Natural History, Chicago.

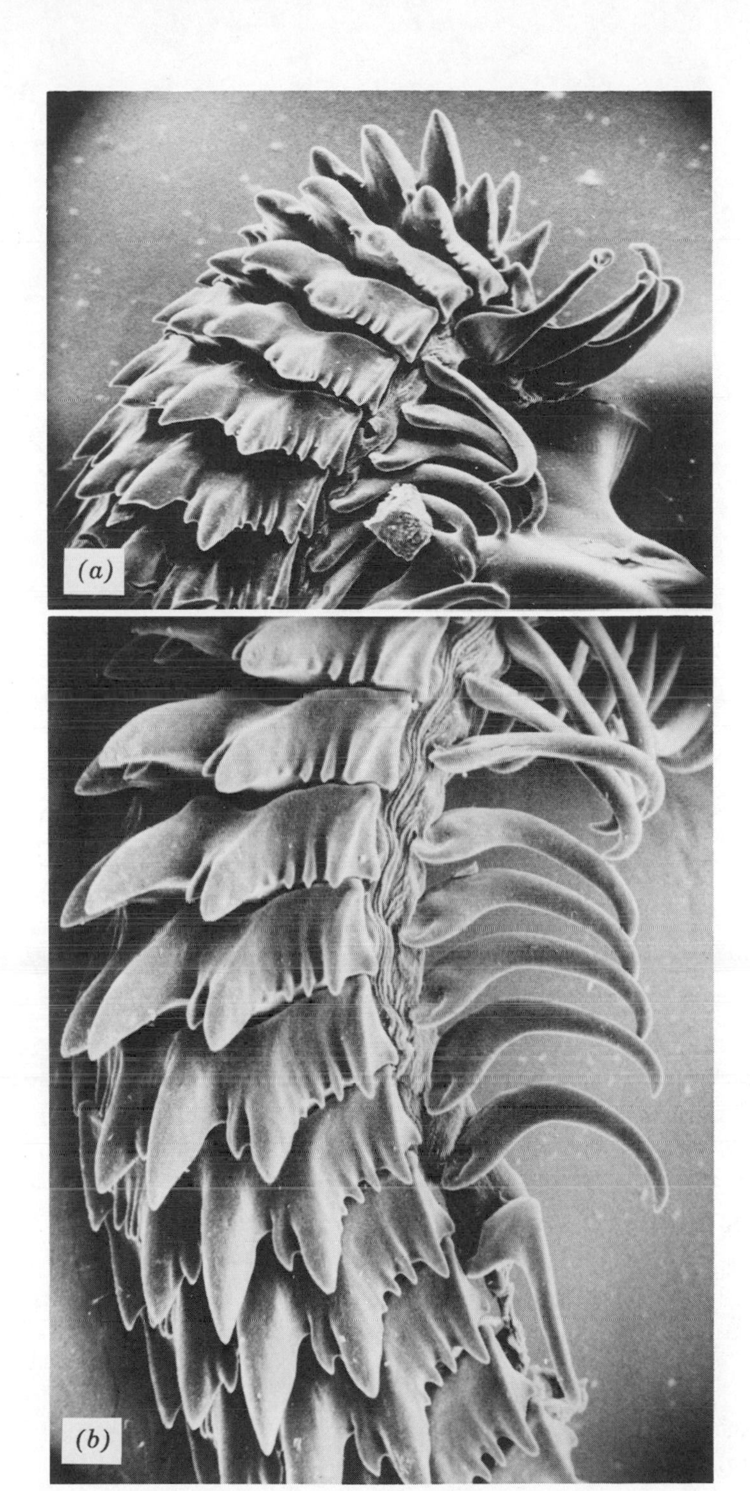
(a)
(b)

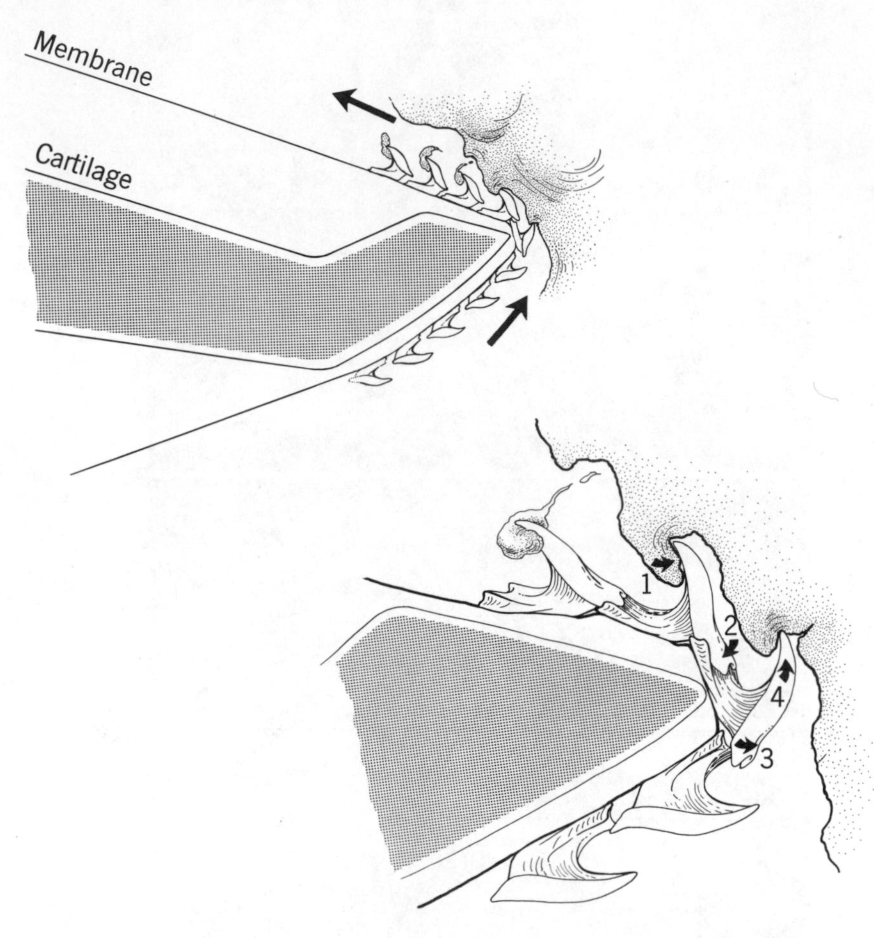

FIGURE 6. Cross-sectional diagram of interrow tooth support during the stress of feeding in pulmonate snails: *upper,* overall pattern of tooth movement and position change at stages in scraping; *lower,* detail of tooth interaction (1–4) at tip of radula.

and twisted teeth are examined (Fig. 3*b* to *d*), then the structure is far clearer. The central tooth (Fig. 3*b*) is V-shaped above, with a series of sharp cusps pointing backward, and small cusps on its base. The lateral tooth (Fig. 3*c*) resembles in shape the capituliform complex of the helicinid (Fig. 2*b*), although the size difference is vast! The larger inner and smaller outer marginals (Fig. 2*d*) of *Gocea* have a rakelike edge with many tiny denticles (pp. 141–142).

Considerable contrast is shown by species of *Ostodes,* a mesogastropod genus from Samoa belonging to the Poteriidae. Seen in relaxed and partly folded position (Fig. 4*a,* p. 145) the teeth seem to have rather small cusps, but this is an artifact of the viewing angle. When not in use these teeth are folded in toward the center (Fig. 4*b*), with the lateral tooth (labeled l) resting on the central (c). The two marginal (m) teeth are folded down on top of the lateral, so that all the teeth lie in a compact mass. The base of the central teeth is notched so that the curved upper portion of the next tooth in line can be "cradled" in this notch. Seen from outside (Fig. 4*c*) the outer marginal is massive and has strong cusps, whereas when the teeth are raised for feeding (Fig. 4*d*) they present a formidable set of points.

In the more primitive stenoglossans, the *rachiglossate* radula has developed with only three teeth in each row, a central with one lateral on each side (Fig. 5, p. 149). In *Drupa ricinus,* a common intertidal muricid in the Indo-Pacific, the central tooth (Fig. 5*a*) is short, with the anterior part of the basal plate (right part of tooth) standing free of the membrane and capable of being supported by resting against the posterior edge of the next tooth in line. In more vertical aspect (Fig. 5*b*), the central tooth is seen as quite wide and with many cusps of different length. In contrast, the right lateral tooth is slender and curved with a hooked tip (Fig. 5*b*). The left lateral tooth is not shown in these views. Presumably the lateral teeth serve to hook and pull into the mouth pieces torn loose by the huge central tooth.

Many publications have illustrated the single harpoonlike tooth of the Conacea. At times reaching a length of more than $\frac{1}{3}$ inch, these formidable spears have a special nomenclature of their own. They represent the acme of specialization in the prosobranch radulae, reduction to a single enlarged tooth connected to a poison gland. Throughout the Prosobranchia there is a clear tendency toward reduction in the number of radular teeth.

The opisthobranchs vary from the single tooth per row characteristic of sacoglossans to forms with over 800 teeth in each row and a total tooth count of perhaps 75,000. It is not possible to make any clear generalizations about their structure.

Pulmonate snails do not show any consistent pattern of change in tooth numbers equivalent to that seen in the prosobranchs. Instead there are multiple changes in cusp structure and tooth number involving food specializations. Carnivores will have fewer and larger teeth than related herbivores. Species that specialize in algal cell piercing and those taking fungal hyphae juices will have extremely numerous and tiny teeth, up to perhaps a maximum of 250,000.

What is common to nearly all herbivorous pulmonates is interactions between the teeth during feeding. The basic type of this interaction is di-

FIGURE 7. Variation in interrow tooth support in three desert camaenid pulmonate snails from Cape Range, Western Australia: (*a*) lateral (upper) and early marginal teeth (lower) in *Globorhagada* showing flexibility of tooth elevation caused by deliberately wrinkled basal membrane, at 500×; (*b*) pattern of interrow overlap in early marginal teeth, at 2000×, on same radula; (*c*) lateral teeth in *Rhagada* showing change in cusp elevation and interlock device, at 960×; (*d*) detail of interlocked lateral teeth on the same radula, at 3940×; (*e*) lateral teeth in a relative of *Pleuroxia* showing reduced basal area of overlap between rows, at 1075×; (*f*) early marginal teeth on the same radula, at 950×, showing that they retain the "flared ridge" interlock seen in *Globorhagada*. Courtesy Field Museum of Natural History, Chicago.

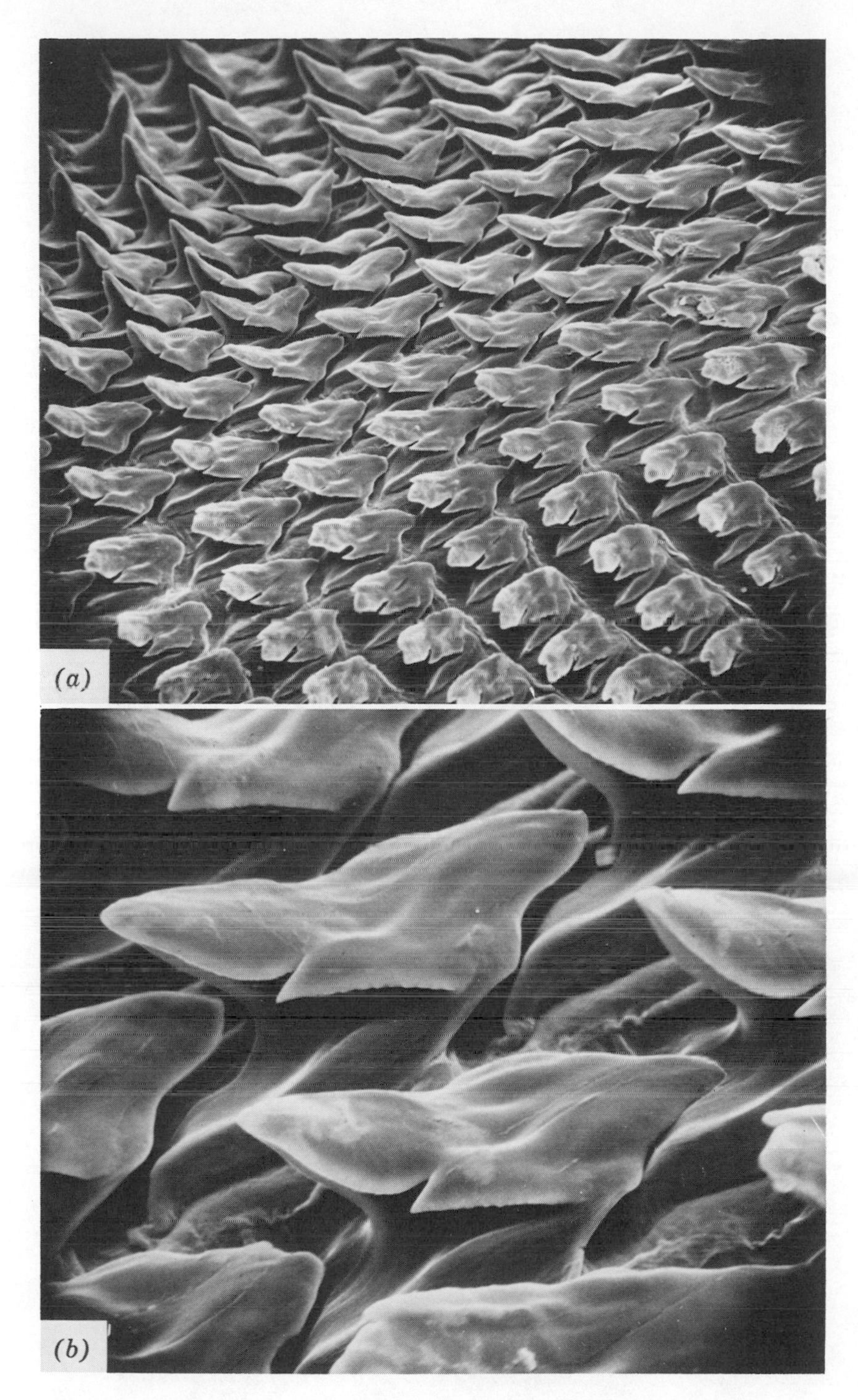
(a)
(b)

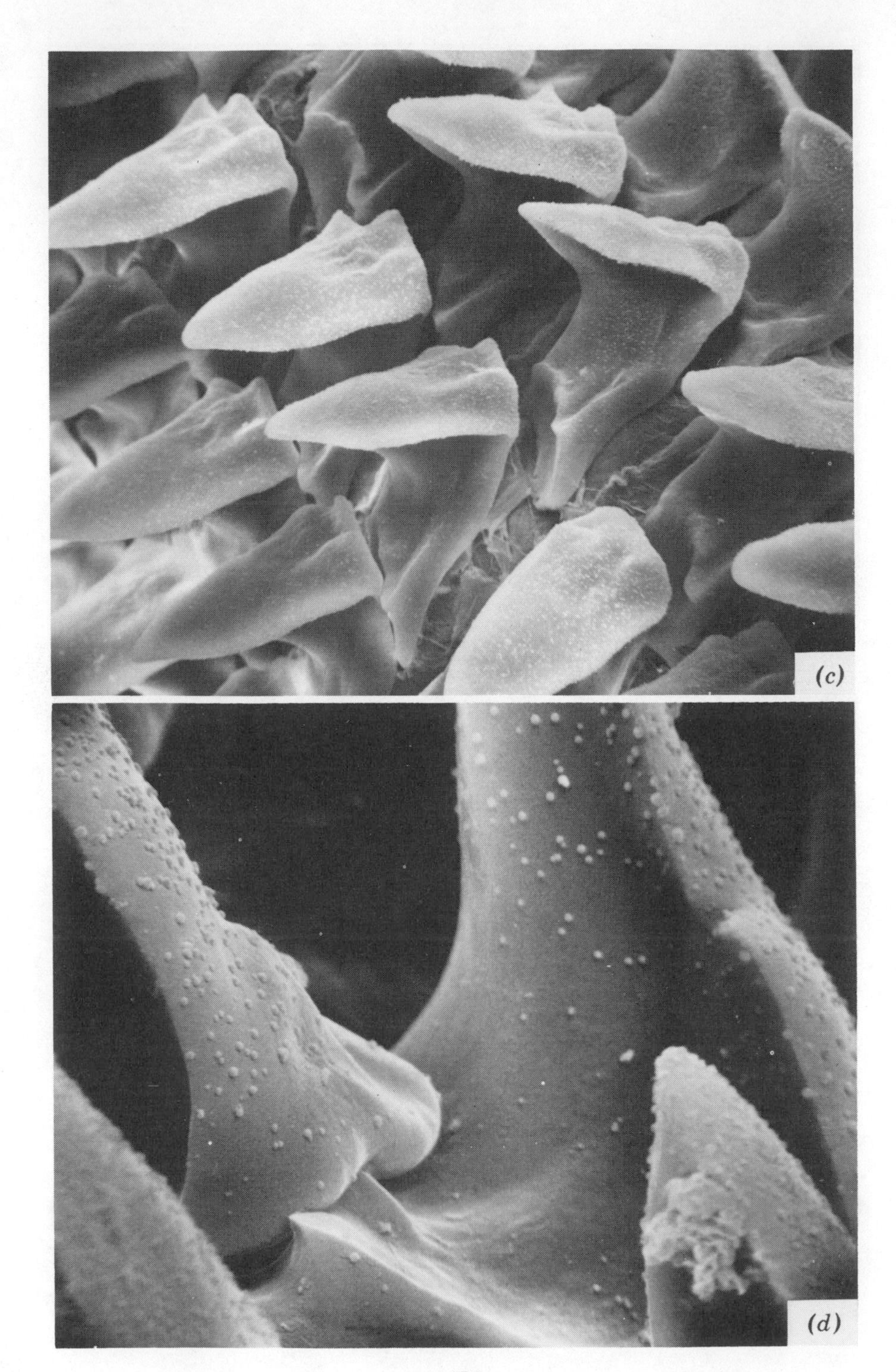
(c)
(d)

(e)
(f)

c
(a)
c
r
r
(b)

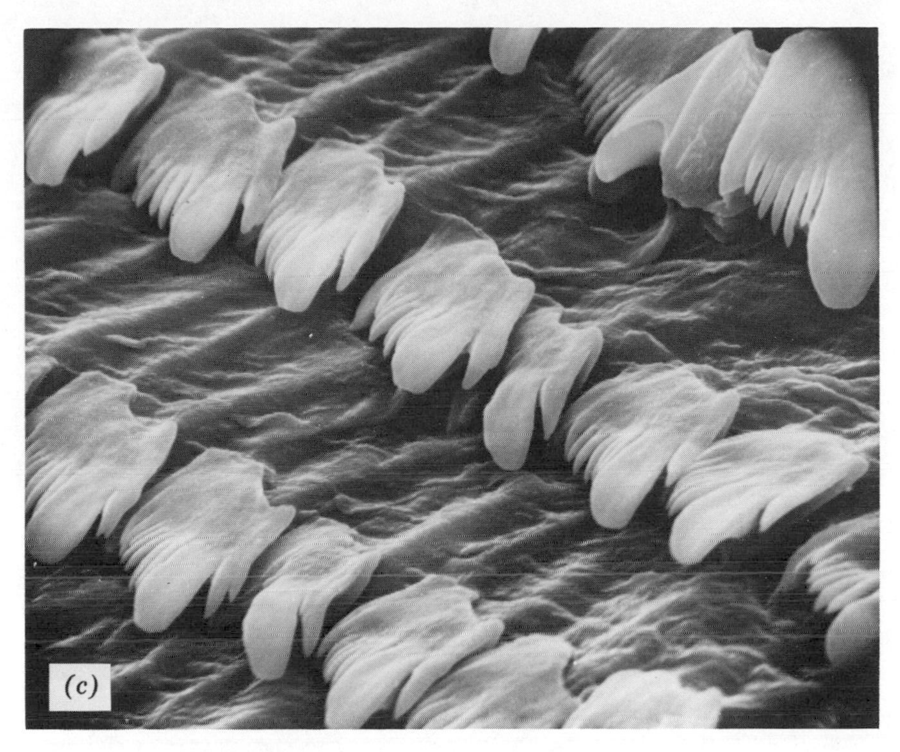

FIGURE 8. Interrow support system in the terrestrial ellobiid *Pythia pachyodon* from Lan-Yi Island off Formosa: (*a*) central (c) and lateral teeth, at 735×, showing transverse support; (*b*) looking posteriorly on central and first lateral teeth showing basal ridges (r) and diagonal support pattern, at 1650×; (c) outermost marginal teeth, at 2950×. Courtesy Field Museum of Natural History, Chicago.

agrammed in Fig. 6. When the underlying cartilages and radular membrane are protruded from the mouth and moved against a food source, the radular membrane is being rotated around the tip of the cartilage (top figure). The elevated cusp of each lateral tooth comes into contact with the food item (bottom figure). Under normal circumstances the cusp will encounter resistance pressure (1) that will force the edge of the cusp backward. This pressure pushes down a flared ridge or other support arrangement at the anterior end of the tooth (remember that the teeth point *backward* into the mouth, so the elevated cusp is *posterior* and the flared ridge is *anterior*). This ridge is pressed downward (2) against the basal plate of the next tooth in line. At the time of maximum resistance encountered by the first tooth,

FIGURE 9. Specialized teeth in land snails: (*a, b*) teeth of *Achatinella bellula* from the Koolau Mountains, Oahu, Hawaii, showing detail of cusp structure at 4975× (*a*), and basal plate length of a displaced tooth at 2575× (*b*); (*c, d*) central and early lateral teeth of *Samoana canalis,* a partulid land snail from the mountains of Upolu Island, Western Samoa, in both vertical view at 1500× (*c*) and laterally at 1150× (*d*). Note the very large overlap between longitudinal rows. Courtesy Field Museum of Natural History, Chicago.

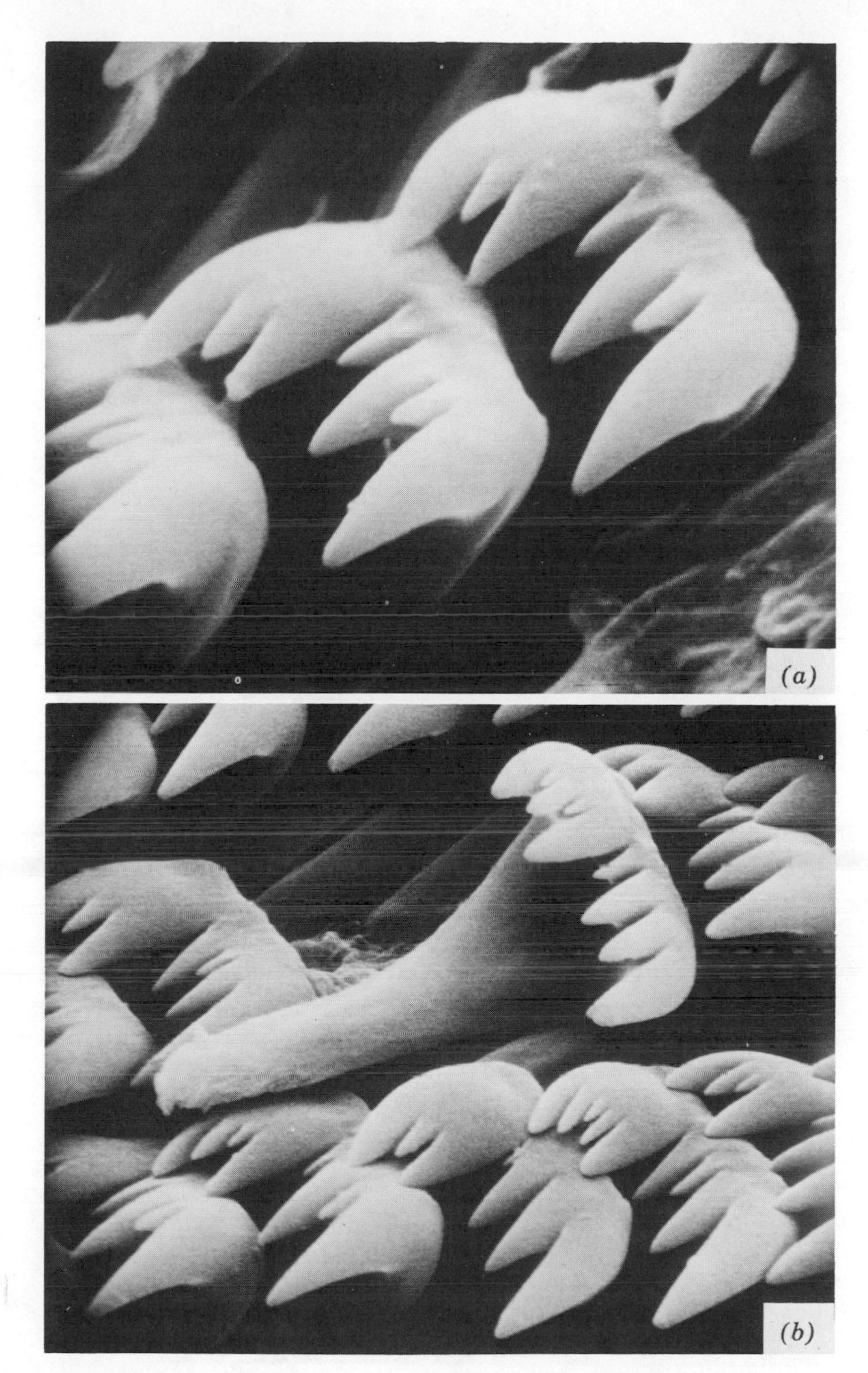
(a)
(b)

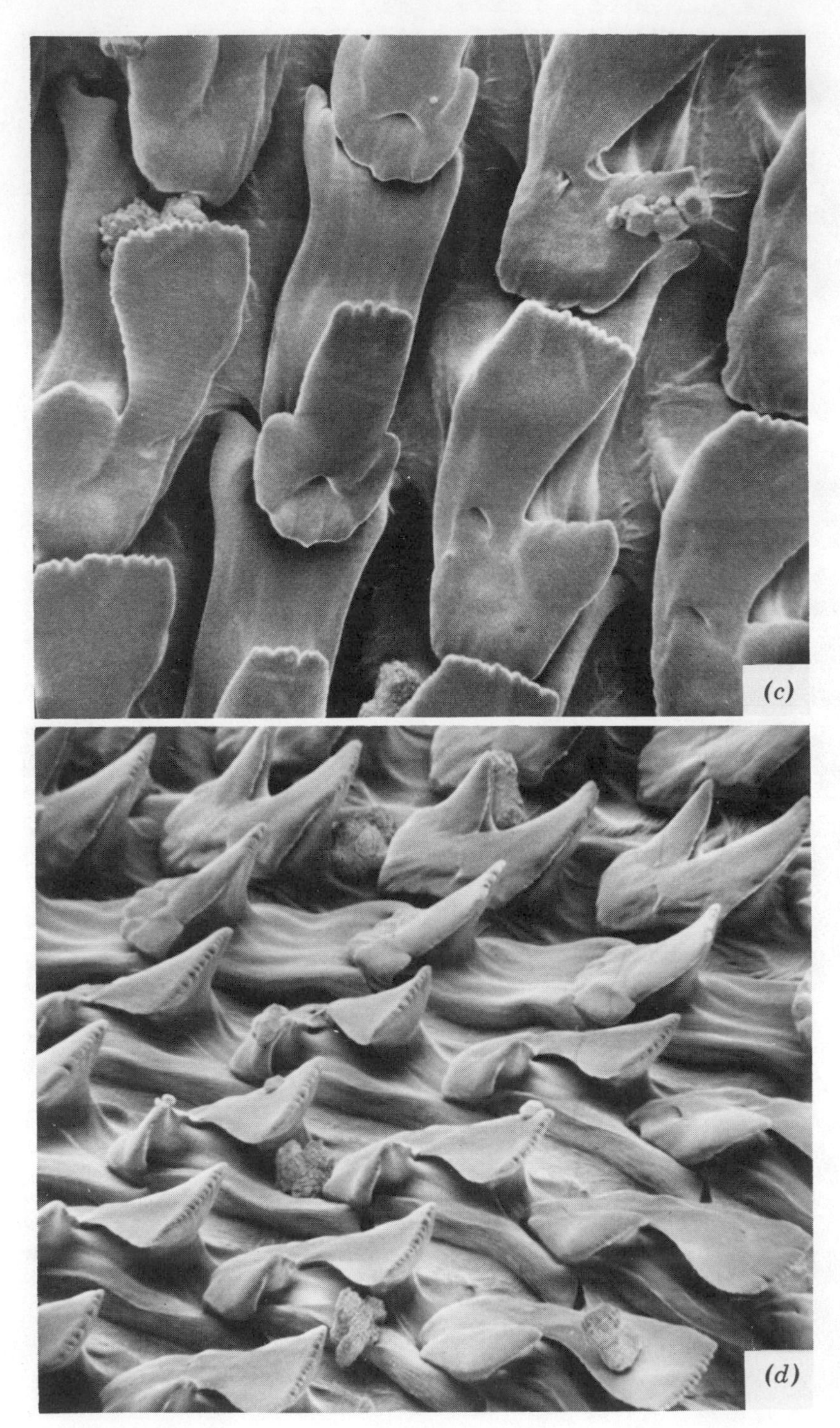
(c)
(d)

the next one in line is rounding the tip of the cartilage and is thus "balanced" like a see-saw. The downward pressure on the basal plate (2) will be translated through the fulcrum of the cartilage tip into an upward thrust (3) for the anterior ridge plus an impetus (4) to the cutting action of the cusp that is just coming into contact with the food item. In this ingenious fashion, resistance encountered by one tooth will be translated through simple lever action into assistance for the cutting action of the next tooth. While it is perhaps too strong a statement to claim that the "food" helps itself be eaten, the pattern of interrow support by the teeth in many pulmonates is a great aid to the eating process. This diagram is simplified in that it shows only one *longitudinal* row in action. Ten to a hundred *transverse* rows may operate in cutting while the marginal teeth in each transverse row are picking up and pulling in torn bits of food (see p. 150).

The exact nature of the interrow support mechanism differs widely among pulmonates and even within the same family. Three species of West Australian desert camaenids probably specialize on different food plants requiring very distinctive patterns of interrow support. *Globorhagada* (Fig. 7*a, b*) has a very large flared ridge that broadly overlaps the basal plate (*b*). *Rhagada* (Fig. 7*c, d*) has a rather narrow overlap, the cusp is narrower and sticks up at a sharper angle, and the interrow support consists (*d*) of a "point and pit" interlock. Yet another species, related to *Pleuroxia* (Fig. 7*e, f*), has the outer cutting teeth (*f*) retaining a flared ridge, while the larger and more central cutting teeth (*e*) have almost no actual overlap, the basal plate of one locking into the cusp edge of the next (pp. 153–155).

Such supports are not restricted to simply the longitudinal rows of teeth, but can shift quite easily from row to row (Fig. 8). In the Ellobiidae, which many think are the most primitive land snails, the general pattern seems to be for a tiny central tooth (c) (Fig. 8*a*) to have a pair of enormous ridges (r) on its basal plate, against which the lateral teeth in the next row on each side can "rest" (Fig. 8*b*). Each lateral tooth has a single ridge on the outer side of its basal plate, against which the cusp of the next outermost lateral tooth can rest. Thus the first lateral on each side is supported by the central in the next row, but the second lateral is supported by the first, and so on, until the marginals (Fig. 8*c*) are reached. These teeth, as most marginals, apparently function only to catch torn pieces. Their low shape, multiple cusps, and wide separation of the rows indicate that no interrow action takes place (pp. 156, 157).

Variations in cusp structure are equally striking and correlate with food source. The Hawaiian tree snails of the Genus *Achatinella* (Plate 12, second from top, bottom) have about 56,000 undifferentiated teeth that closely overlap (Fig. 9*a, b*). A long slender basal plate curves upward to a short

FIGURE 10. Carnivorous land snail teeth: (*a, b*) *Euglandina rosea* from Florida showing the teeth in folded (*a*) and feeding (*b*) positions. The arrow in *b* indicates the anterior supporting ridge. Magnifications are 350× for *a* and 1000× for *b*; (*c, d*) *Ptychorhytida aulacospira* from New Caledonia, South Pacific showing a partial row of elevated teeth (*c*) viewed at 230× from the center of the radula and a whole row of folded teeth (*d*) at 230×. Courtesy Field Museum of Natural History, Chicago.

(a)
(b)

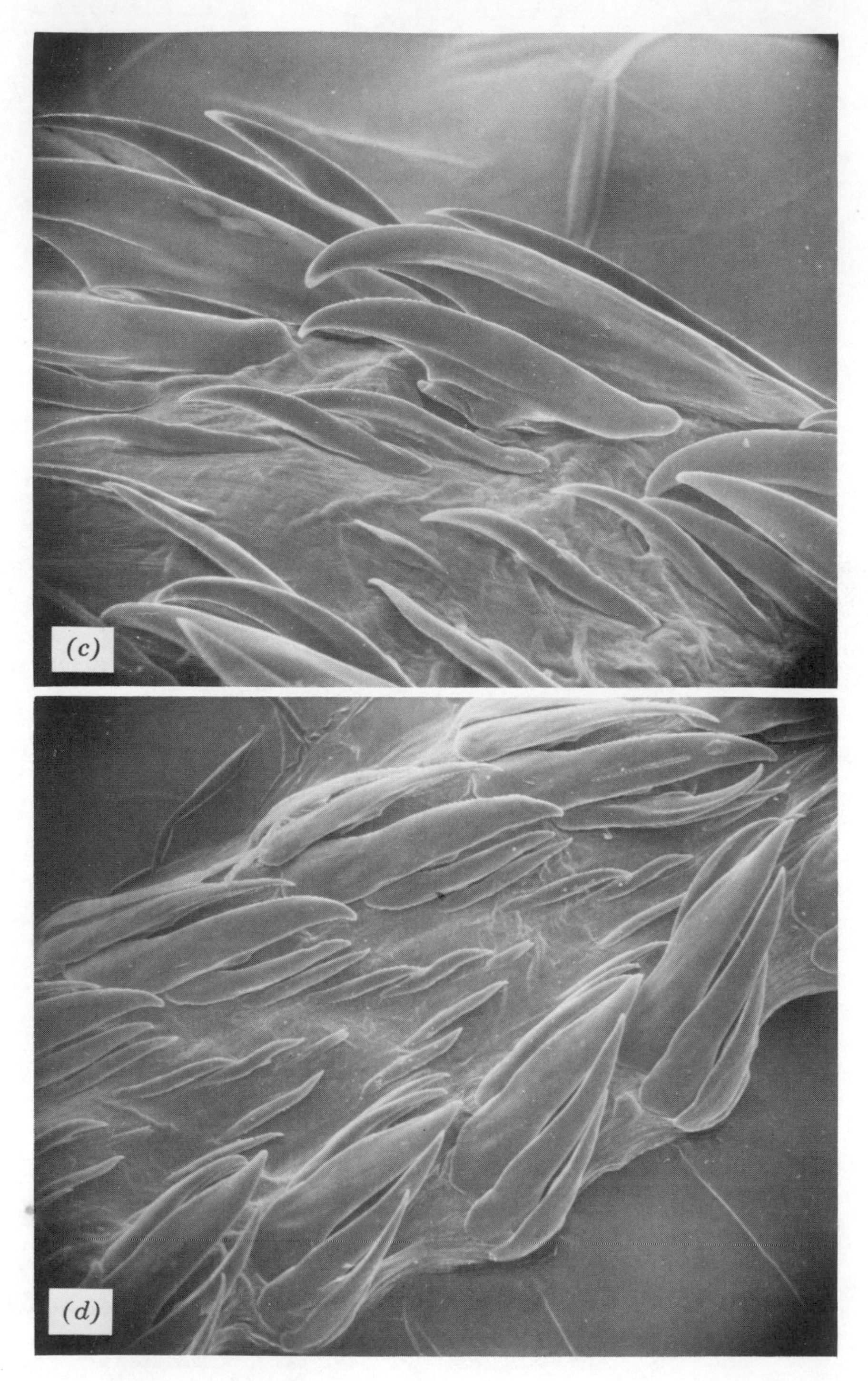
(c)
(d)

extension with 8 to 10 cusps that point directly forward. These are extremely tiny teeth that apparently puncture and partly scrape loose the cells of algal films found on the surfaces of leaves and twigs. A very different pattern is shown by the teeth of *Samoana,* which is also a tree dwelling group found on high islands in the South Pacific. Seen from directly above (Fig. 9*c*) the central tooth has a single broad spadelike cusp, while the laterals on each side show a larger and a smaller cusp. In side view (Fig. 9*d*) the cusps are seen to have a serrated edge for scraping. The two cusps would be used at a different time: The larger, straight-edged cusp would scrape across the surface first, then the smaller, curved cusp would catch and pull loosened materials. The same spadelike cusp is seen in *Papuina* (Fig. 1*c, d*), and is also found in several other unrelated families.

Carnivorous species have the number of teeth greatly reduced, ranging from less than a dozen teeth in a row to perhaps 40 or 50 teeth in a row. Because their teeth are so long and sharp, they must be compactly folded when not in use and strongly elevated when functioning. This is achieved in different ways by different snails. The Neotropical genus *Euglandina* (Fig. 10*a, b* and Chapter XIII, Fig. 4*b*) has the teeth folded and overlapping when not in use (*a*), but when raised for slicing (*b*) the anterior edge is seen to be truncated (marked by an arrow) and thus can be supported by the underlying cartilage when pressured by a slicing action. The New Caledonian paryphantid *Ptychornytida* is quite different. Here the smallest teeth are in the center of the radula (Fig. 10*d*), with the largest near the outer edges. The large teeth can be folded over on top of each other to be out of the way, but when in a functioning position (Fig. 10*c*) the notched inner edges that allow this compact folding are evident. The low angle and great length of the teeth indicate that *Ptychorhytida* feeds with a "stabbing stroke" rather than the "slicing stroke" of the *Euglandina.* Since carnivores have evolved in many different groups of land snails, the differences in radular functioning are to be expected.

In a very few situations, the differences in cusp structures characterize major phyletic units. Among the advanced land snails, the pulmonate superfamilies Arionacea and Limacacea include 2000 to 4000 species each. The more primitive arionaceans (Fig. 11*a, b*) usually have a tricuspid pattern to the teeth. The central and lateral teeth (*a*) are slenderer and tend to be curved much more strongly (see also *Suteria* in Fig. 1*a*); the marginals (Fig. 11*b*) have short basal plates and often only the three cusps. They point upward and lean very slightly, if at all, toward the center. Generally a transverse row of teeth may number only 40 to 80 teeth, with the total radula containing 3000 to 4000 teeth. In contrast, a typical limacacean, the Rarotonga, Cook Island *Diastole conula* (Fig. 11*c, d*) has about 19,000

FIGURE 11. Tooth structure in arionacean and limacacean pulmonate land snails: (*a*) central (c) and lateral (l) teeth of the arionacean snail *Maoriconcha oconnori* from South Island, New Zealand, at 2150×; (*b*) marginal teeth from the same radula, at 2200×; (*c*) partial row of radular teeth in the limacacean *Diastole conula* from Rarotonga, Cook Islands, South Pacific, at 1150×, with central and lateral teeth in the upper right and marginal teeth on the lower half; (*d*) detail of marginal teeth overlap, at 3050×, on right side of radula; (*e*) central (c) and lateral (l) teeth of the limacacean *Fanulum testudo* from Mt. Bates, Norfolk Island, South Pacific, at 1950×; (*f*) outermost marginal teeth from the same radula, at 2075×. Courtesy Field Museum of Natural History, Chicago.

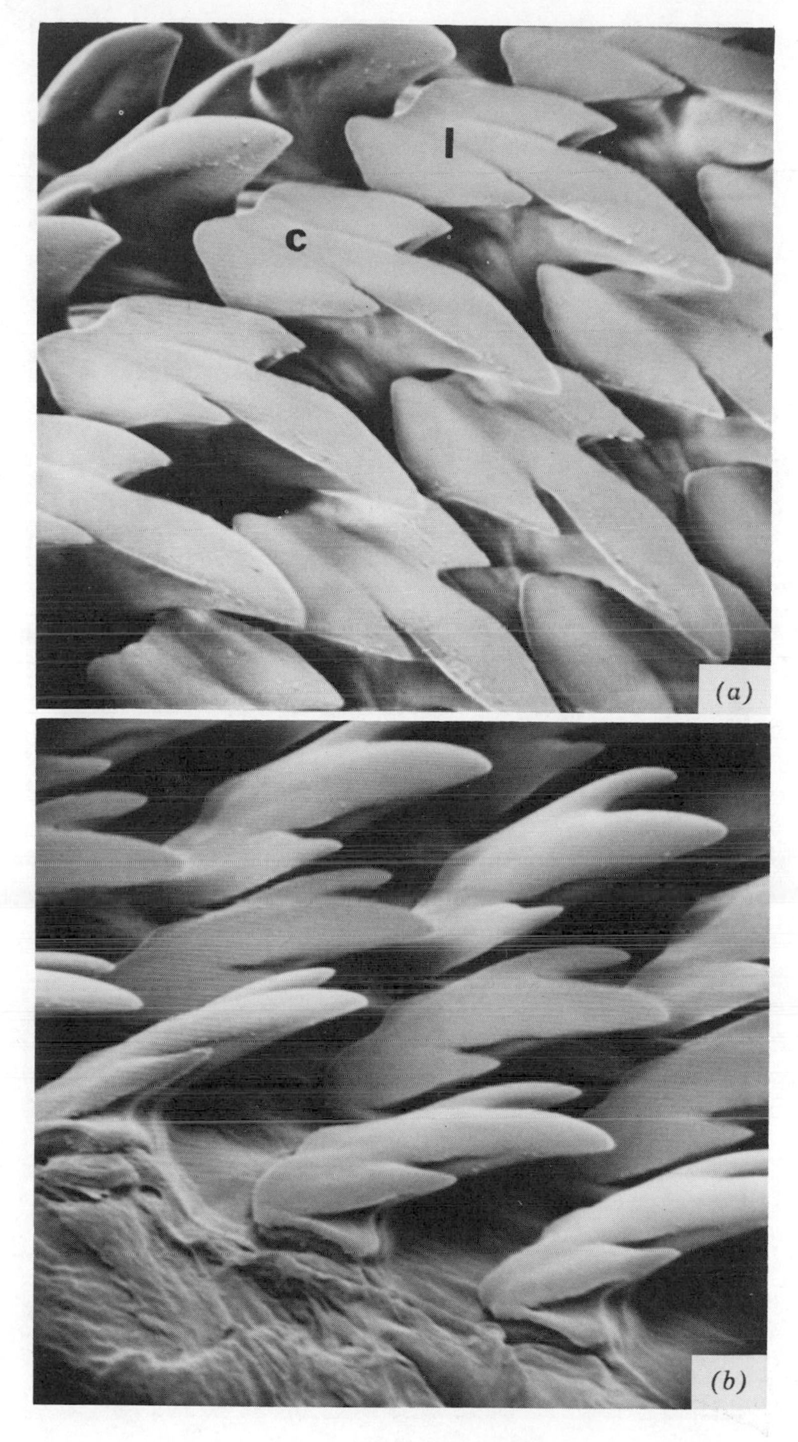
l
c
(a)
(b)

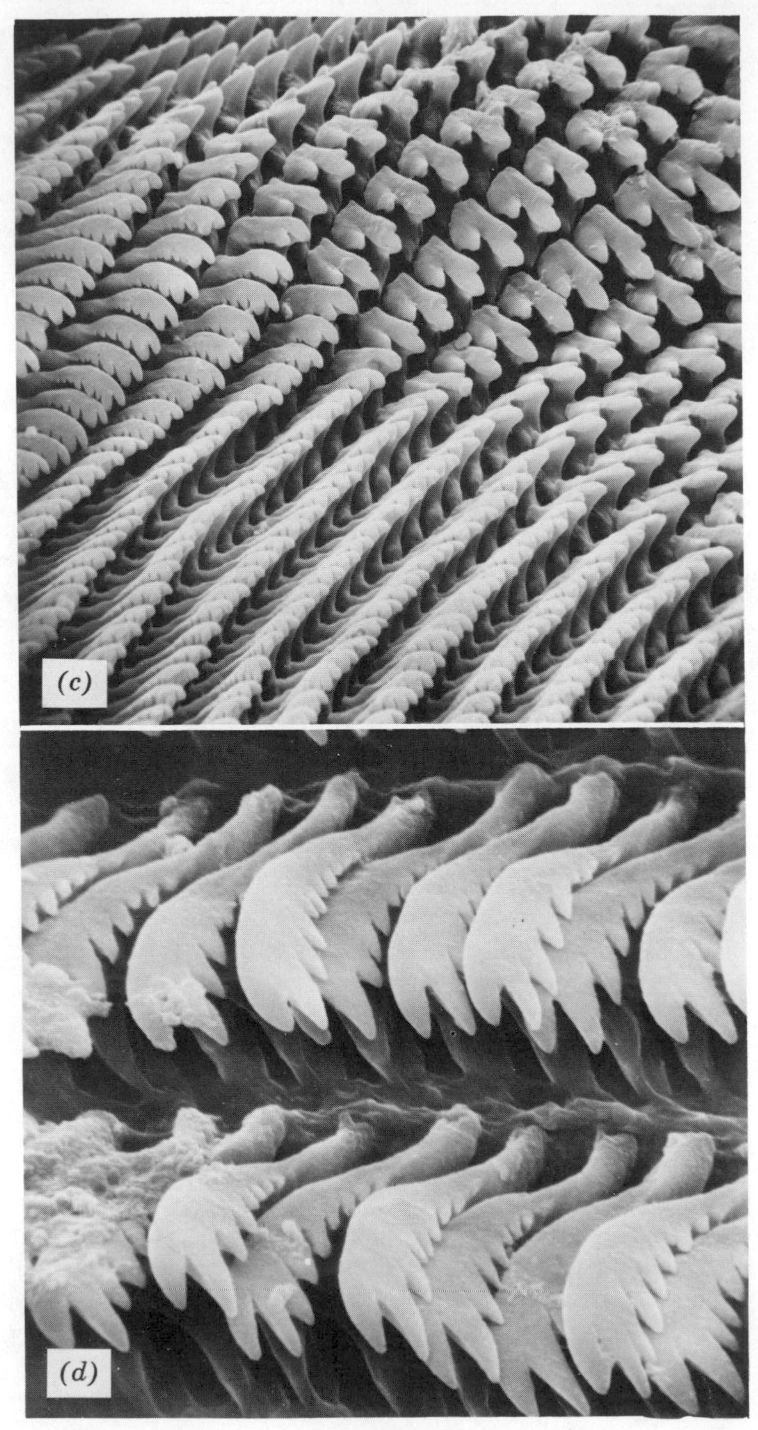
(c)
(d)

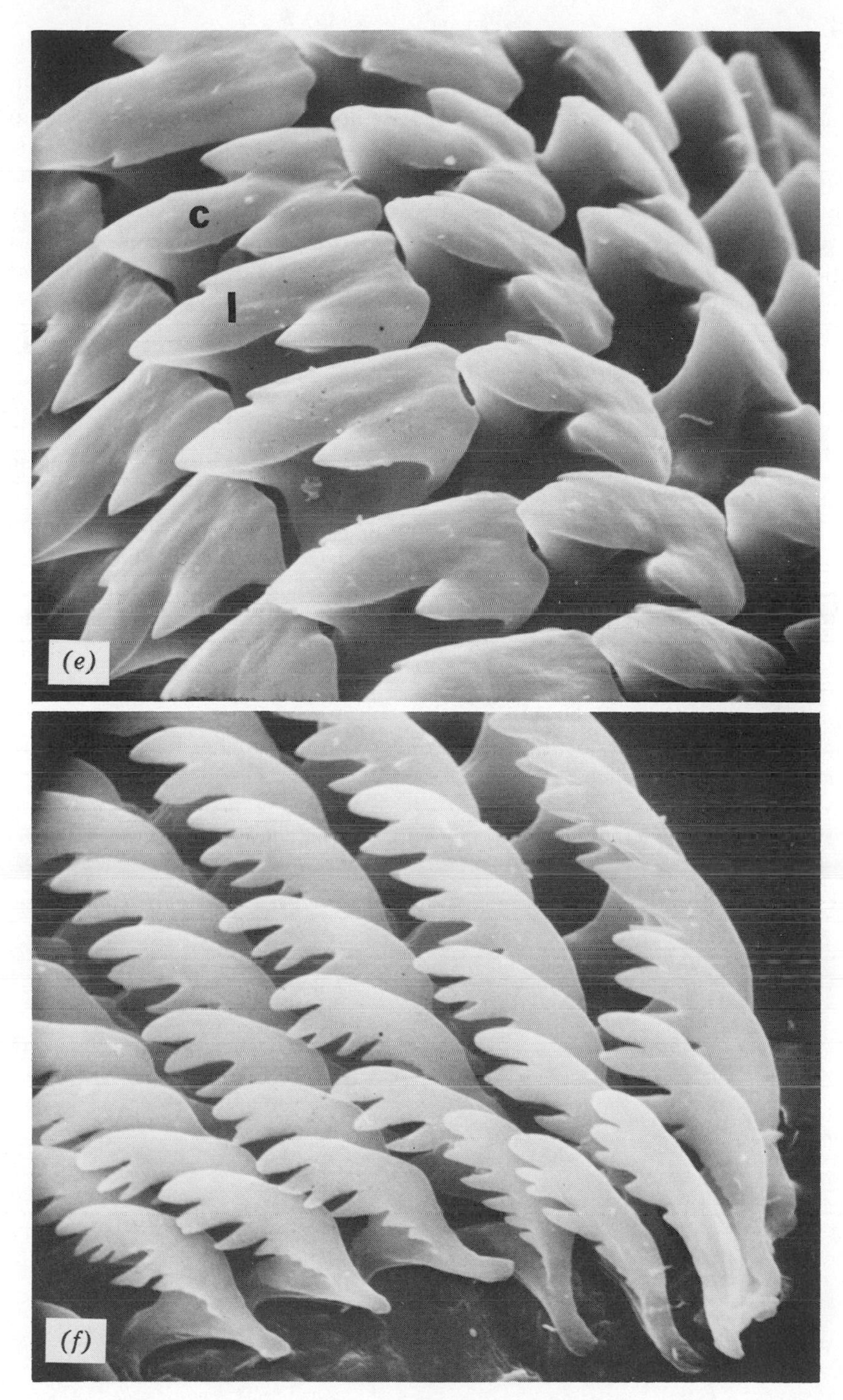
c
l
(e)
(f)

teeth, 181 in each of about 106 rows. There are eight laterals (those with unequal side cusps) on each side of the stubby central tooth (single tooth with equal side cusps in upper right corner of Fig. 11*c*) and about 82 marginals on each side. The marginals (Fig. 11*d*) have several cusps on the outer side of the tooth, long and narrow basal plates, and "lean" distinctly toward the center of the radula. If viewed from nearly the same angle as the arionacean, a radula from a *Fanulum testudo* taken on Norfolk Island in the South Pacific (Fig. 11*e, f*) shows that the limacaceans have much shorter, chunkier central and lateral teeth while the marginals (Fig. 11*f*) are dramatically different in shape.

These descriptions give only a glimpse of the variations shown by the more than 25,000 species of nonmarine mollusks. Evolutionary changes in the patterns of cusp and support structure are obviously one of the prime ways in which snails specialize within local areas or exploit different levels of food resources. The generalized land snails feed on decaying vegetation or fungal hyphae and have mostly tricuspid teeth, although many groups show a tendency toward becoming unicuspid. The specializations for spade like scraping (*Samoana,* Fig. 9*c, d*), algal cell piercing (*Achatinella,* Fig. 9*a, b*), and carnivorous feeding (Fig. 10) are secondary to the tricuspid pattern seen in the majority of species.

These types of radular variations are but one aspect of exploiting life on land. What the basic adjustments were and something of the other experiments in land life are explored in the next two chapters.

X

The Road to Land

Several years ago on a cold day in February, a small package of land snails arrived in my office. They had been mailed from Cairo, Egypt late in July, had sat for several weeks on a steaming dock in Alexandria, been transferred from ship to ship, endured the icy cold and churning waves of a North Atlantic January crossing, and finally met the tender mercies of the United States Parcel Post during a subzero cold wave. Despite these cumulative catastrophies, more than 90% of the snails had survived. All they needed to start crawling happily about was a piece of wet newspaper dumped in their box.

Remarkable as this may sound, it is far from a record. Two specimens of the same species (*Eremina desertorum*) were glued to a piece of cardboard and placed in the mollusk collection of the British Museum (Natural History) on March 25, 1846. Almost four years later, about March 15, 1850, the cardboard was placed in some water, the glue dissolved, and one snail crawled away. It had survived four years without any food, exercise, or water. Some Mexican desert snails, *Micrarionta veitchii* and *Orthotomium pallidior,* set an even longer record. They lived in a box for six years hidden in the back corner of a desk drawer belonging to a San Francisco scientist.

These are only extreme examples of one solution to major difficulties of living on land. Climates and weather vary. Any land dweller must face periods of time when conditions are not favorable. Most English-speaking people live in areas where a warm season (summer) is balanced against a cold season (winter). But much of the world has an alternation of *wet* and *dry* seasons. In cold weather water is difficult for animals to get because it is frozen into ice. In dry seasons there is little or no rain, the ground and ponds dry up, and water is just not available. Life originated in water and land organisms must get water from somewhere in order to live.

How can organisms survive these bad times when water is not available? In different ways. Birds may migrate distances of hundreds or thousands of miles to escape winter. Trees and shrubs will shed their leaves and become dormant. Grasses and other perennial plants die back, with the roots producing new shoots when warm and wet spring weather returns. Many insects lay eggs that winter over, while butterflies and moths become pupae locked in silken cocoons. Bats and bears hibernate in secret hollows or caves. They fall into a deep sleep, their heartbeat slows, and their temperature falls to within a degree or so of freezing. In this state of "suspended animation" they survive the cold winters.

During dry seasons animals go into a state equivalent to hibernation called *aestivation.* Their heart beat slows, and they become totally inactive. Frogs and toads can become sealed in dry mud. The African aardvark aestivates in its burrow. Microscopic rotifers dry out and shrivel up, to be brought back to life when the next film of water wets the ground.

The snails retreat into their shells.

Having a shell on their back, a mobile garage into which the body can be withdrawn, provided the means for a snail to venture forth onto land. In the ocean, pulling the head and body into the shell and shutting off the world with an operculum was chiefly a means of defense against being eaten. On rocky shores where there is the ebb and flow of tides, the retreat into a shell provided as much protection against drying out as against being eaten.

We do not know when the first snail wandered inland from the tidal zone, or even what route it took. It might have been along the banks of a river or through a swamp. It could have been directly from the seashore into a nearby patch of mosslike plants. Certainly it must have been very shortly after the first land plants evolved. As soon as there were plants on land, there were dead plant bodies for snails to feed on. The early plants lived under very wet conditions, and the first snails undoubtedly crawled in moist comfort.

Temporary dryness in the heat of midday or between showers was no different than being stranded at high tide. The snail simply retreated into its shell to wait for wetness. Many times the wait would have been too long and the snail died and dried out, still retracted within its shell. Even today when collecting in the limestone hills of Cuba or Malaya during the dry season, it is far more common to find shells with operculum in place and only a dried remnant of the animal inside than living specimens.

But a few hardier snails survived each drought. Their progeny explored different ways of living and evolved. After 350,000,000 years of living on land their descendants number more than 24,000 species in some 1300 genera belonging to 70 different families. Vertebrate land animals are fewer in number—perhaps 3500 species of mammals, 8600 kinds of birds, about 6000 different reptiles, and only 2500 amphibians for a total of 20,600 species. Although land snail diversity pales in contrast with the more than 1,000,000 known species of insects, spiders, mites, and other land arthropods, mollusks are the second largest group of land animals and show an amazing variety of structures and habits.

The earliest land snails were found in fossil trees, together with various insects and early terrestrial vertebrates. The specimens have been taken from coal beds in New Brunswick; near Joggins, Nova Scotia; near Marietta, Ohio; and localities in Vermilion County, Illinois. Ranging in age from Mississippian to Lower Permian, or 350,000,000 to 260,000,000 years ago, they show a variety of shape and form that has caused scientists to argue for more than 100 years as to their relationships. These arguments still continue. We can only say that a variety of land snails lived that long ago.

The largest species discovered, *Dendropupa vetusta* (lower row and upper left of Fig. 1), is about $\frac{1}{3}$ inch long, has nine whorls, is cylindrical in form, and has a sculpture of radial striae. The smallest species, *"Zonites" priscus* (upper right of Fig. 1), is $\frac{1}{10}$ inch in length, has a sculpture of both strong and fine radial ribs, and has four whorls coiled in a typical helicoid fashion. In contrast, the $\frac{1}{8}$ to $\frac{1}{6}$ inch *Maturipupa* and *Anthracopupa* (Fig. 2) are cylindric-conic in form, have only four to five whorls, one or more barriers in the shell aperture, and fine radial sculpture. *Dawsonella meeki* (Fig. 3) has a huge umbilical callus and a double-edged lip, no prominent surface sculpture, and only about $3\frac{1}{4}$ to $3\frac{1}{2}$ whorls.

No fossil land snails are known from these Paleozoic fossils up to Paleocene times, about 70,000,000 years ago. The many snails found in the Paleocene are readily referred to modern families. The gap of 190,000,000 years from the Lower Permian period of *Anthracopupa* to the Paleocene of modern land snail families conceals most of land snail evolution. We have no fossil record of the changes, but we can learn something of this evolution by studying living land snails. By comparing structures, patterns of behavior, distribu-

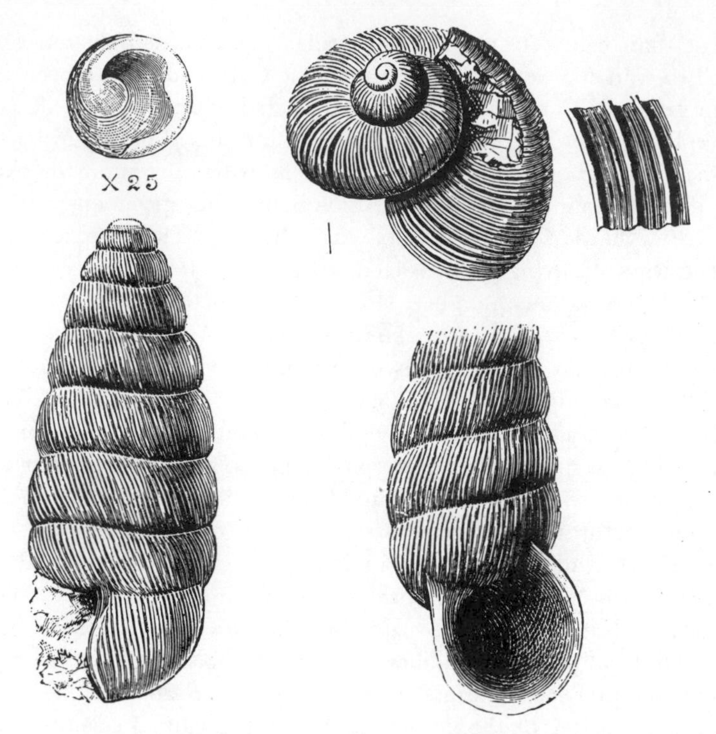

FIGURE 1. The first known land snails. Found in Carboniferous deposits of Eastern Canada. Upper right: *"Zonites" priscus* with a detail of its shell sculpture; upper left and lower row: embryo shell and two views of adult *Dendropupa vetusta*. Taken from an 1880s Christmas card sent by Dawson and based on his published work.

tion, and differences in function, it is possible to trace the major changes in land snails and to understand why they are such a successful and diverse group today.

The pulmonates and the prosobranchs are the two groups of snails that migrated onto land. The prosobranchs are derived directly from families of marine snails. They survived by changes in patterns of behavior more than by changes in structures. The pulmonates on the other hand, show major changes in structure from those of their marine ancestors. Plate 10 shows a typical land prosobranch, *Ostodes tiara* (upper left) from Upolu, Western Samoa in the South Pacific, and an advanced pulmonate, *Meridolum gulosum* (upper right), that lives near Sydney, Australia. As do nearly all prosobranchs, *Ostodes* has an operculum sitting on the tail, the black eye spots are at the base of the tentacles, and the mantle forms

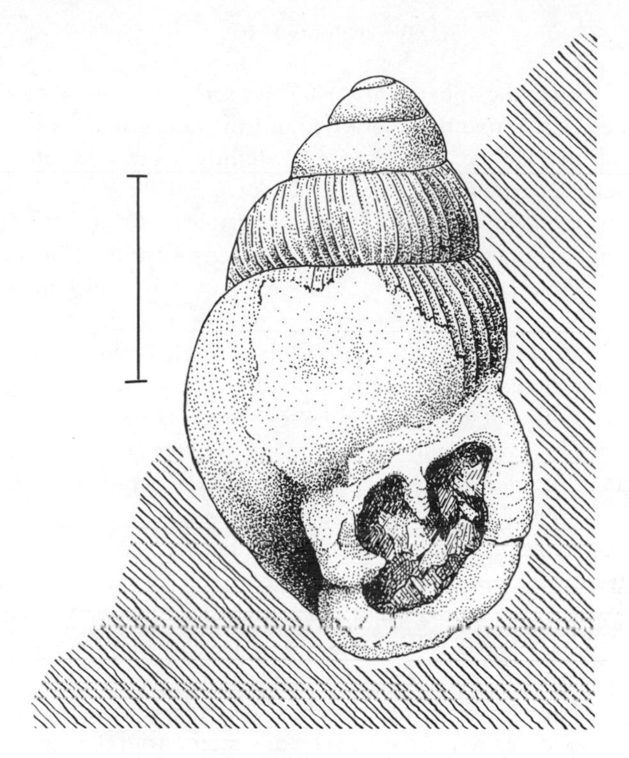

FIGURE 2. *Anthracopupa ohioensis,* a fossil land snail from the Carboniferous of Belmont County, Ohio. Scale line equals 1 mm.

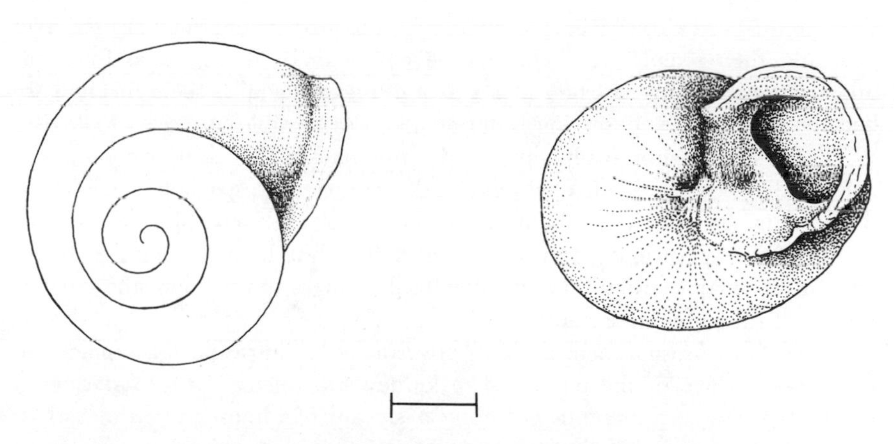

FIGURE 3. *Dawsonella meeki,* a fossil land snail from the Carboniferous of Vermilion County, Illinois. Scale line equals 1 mm.

a free collar lining the inner margin of the shell. The neck of the animal is not fastened to the mantle anteriorly, and the entire mantle cavity is open to the flow of air. The head of *Ostodes* is slightly protruded into a proboscis and is distinctly separated from the broad foot.

Meridolum shows a number of obvious external contrasts. It lacks an operculum on its tail, the eye spots sit on the tip of much longer tentacles, which can be retracted into the body; there is a smaller, second set of tentacles developed; and instead of there being an open mantle cavity, the mantle collar is fused to the neck of the snail. Only a single circular hole, the pneumostome, connects the mantle cavity to the outside world. The pneumostome is usually surrounded by a set of muscles that can narrow or expand the opening, depending on whether the snail is actively breathing or trying to conserve water. In addition, the head of *Meridolum* is only slightly separated from the foot and is not prolonged into a proboscis.

Land prosobranchs number about 3650 species, compared with 20,500 species of terrestrial pulmonates. Prosobranchs are confined mainly to the wet tropics, while pulmonates occur from the islands of the Canadian Arctic and Siberia to some of the driest deserts as well as moist forests and grasslands. Pulmonates are more successful in living on land and occupy a much wider array of habitats. They have evolved mechanisms for conserving water and are able to carry a reserve supply of water around with them, while the prosobranchs continue to depend on retreating into the shell and sealing the opening with an operculum.

In both groups some basic changes from life in water were required. Respiration could no longer take place by gills in the mantle cavity exchanging gases with water swept across their surface. Undigested food and excretion products could not be picked up by the water that had crossed the gills and then swept out of the mantle cavity to be dispersed as food for other organisms. Instead of water surrounding the snails, their life had to become adjusted to conserving water as much as possible. A Welsh zoologist, W. J. Rees, has whimsically christened land snails as ". . . hypochondriacs perpetually conscious of environmental conditions, especially moisture." Books have been written about the biochemical and physiological adjustments made by various organisms to land life. But here I intend to focus only on the major structural and behavioral changes that explain the greater success of the pulmonate snails.

In the land prosobranch, the gills are lost and respiration takes place on the inner surface of the mantle. The kidney still opens at the posterior of the pallial cavity, apparently secreting a stream of almost pure water that moistens the surface of the pallial cavity. This aids respiration and keeps the tissue from drying out. Undigested food is passed out the right anterior margin of the pallial cavity. In the case of the European land prosobranch, *Pomatias elegans,* experiments have shown that it will be active only when the ground level humidity is 95% or more, indicating the marginal efficiency of this arrangement. In a period of high humidity, during or just after a heavy shower, the prosobranch can crawl about, feed, and do all the necessary activities of living.

But in less than optimum wetness, all the snail can do is retreat behind the operculum and wait for wet weather to return. In the rain forest areas of the world, where it is wet all the time and exceedingly wet some of the time, the next shower will come in a day or so. But there are many, many tropical areas that have a monsoon climate, where part of the year is very dry, and part exceedingly wet. One of the greatest centers of diversity for prosobranch land snails is in the Khasi Hills of Assam, northeastern India. Here the winter rainfall may total less than 1 inch a month (about the average rainfall of Salt Lake City, Utah), while June and July have a total of 200 inches. In these two months the prosobranch land snails may be washed away by floods, but in the winter months there will be almost no time when they can be active.

If the operculum seals the aperture so tightly that water loss will be prevented, how can the aestivating snail breathe? It must continue extracting oxygen from the air and getting rid of carbon dioxide. The means of doing this has been adopted independently by land prosobranchs in such widely separated areas as India, Australia, Malaya, Cuba, and Mexico. Various notches, grooves, slits, or separate tubes provide space for air to circulate between the outside and the pallial cavity even though the snail is retracted. Several examples are illustrated in Figs. 4 and 5. *Aperostoma mexicanum palmeri* from San Luis Potosi, Mexico (Fig. 4*a*) has the last part of the body whorl detached. A deep circular notch in the parietal wall allows breathing. While the animal can retract deeper into the shell than this notch, it can and does stay with the operculum blocking the aperture, but with the notch open to the pallial cavity. *Pterocyclos prestoni* from North Vietnam (Fig. 4*b*) shows a groove in the upper lip that performs exactly the same function. From this groove to the closed tube seen in the Cambodian *Rhiostoma hainesi* (Fig. 4*c*) is simply a matter of continuing growth a little further and closing the outer side of the groove. This particular strategy has been carried to its logical conclusion in the Malayan *Rhaphaulus lorraini* (Fig. 5*a*). A groove in the upper margin of the aperture extends posteriorly for almost one-half whorl. About one-eighth whorl inside the aperture it becomes a closed tube that is turned upward on the outside. A different method is used by the Queensland pupinid species *Pupinella simplex* (Fig. 5*b*). A notched groove on the columellar wall extends quite a distance back-

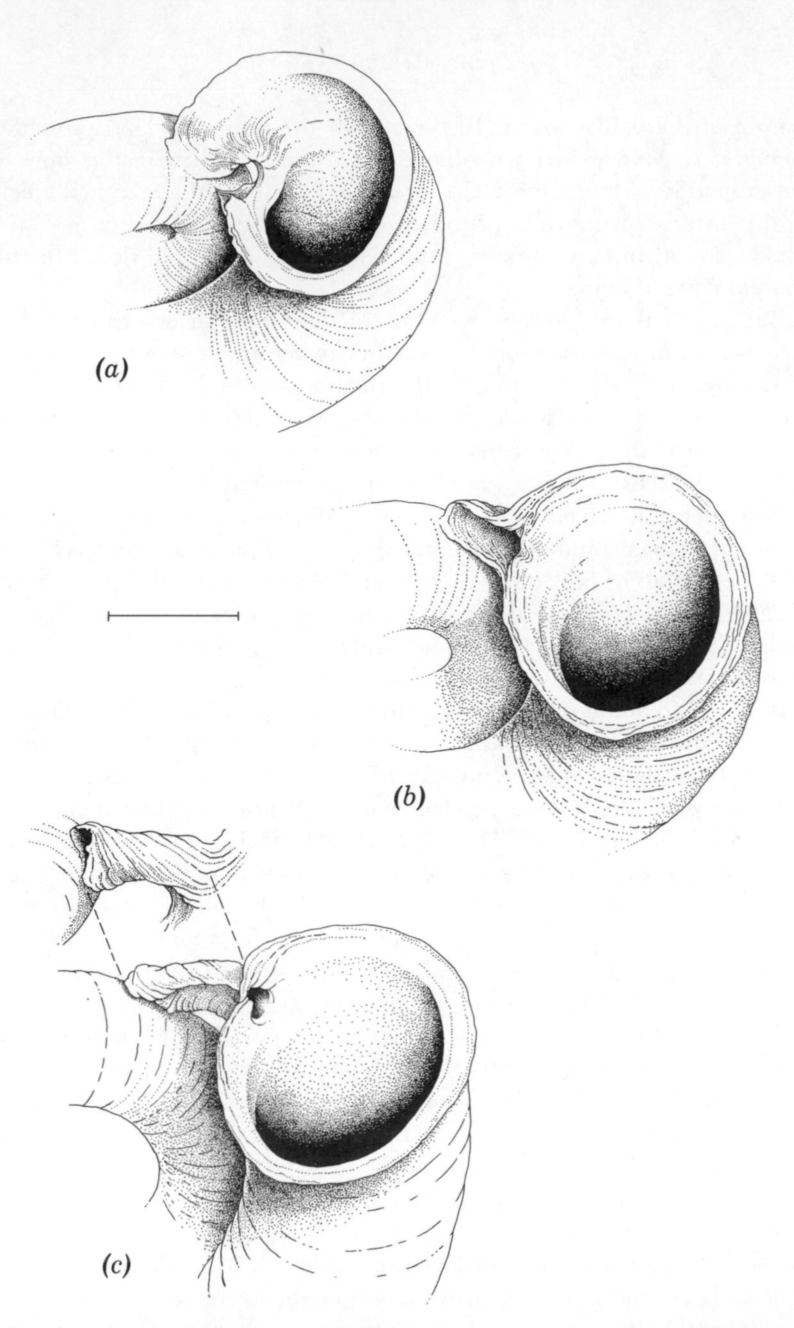

FIGURE 4. Breathing slits in prosobranch land snails: (*a*) *Aperostoma mexicanum palmeri* from Xilitla, San Luis Potosi, Mexico; (*b*) *Pterocyclos prestoni* from Muong Hum, Tonkin, Vietnam; (*c*) *Rhiostoma hainesi* from Cambodia. The detailed drawing shows the breathing slit as seen from the upper side of the shell. Scale line equals 5 mm.

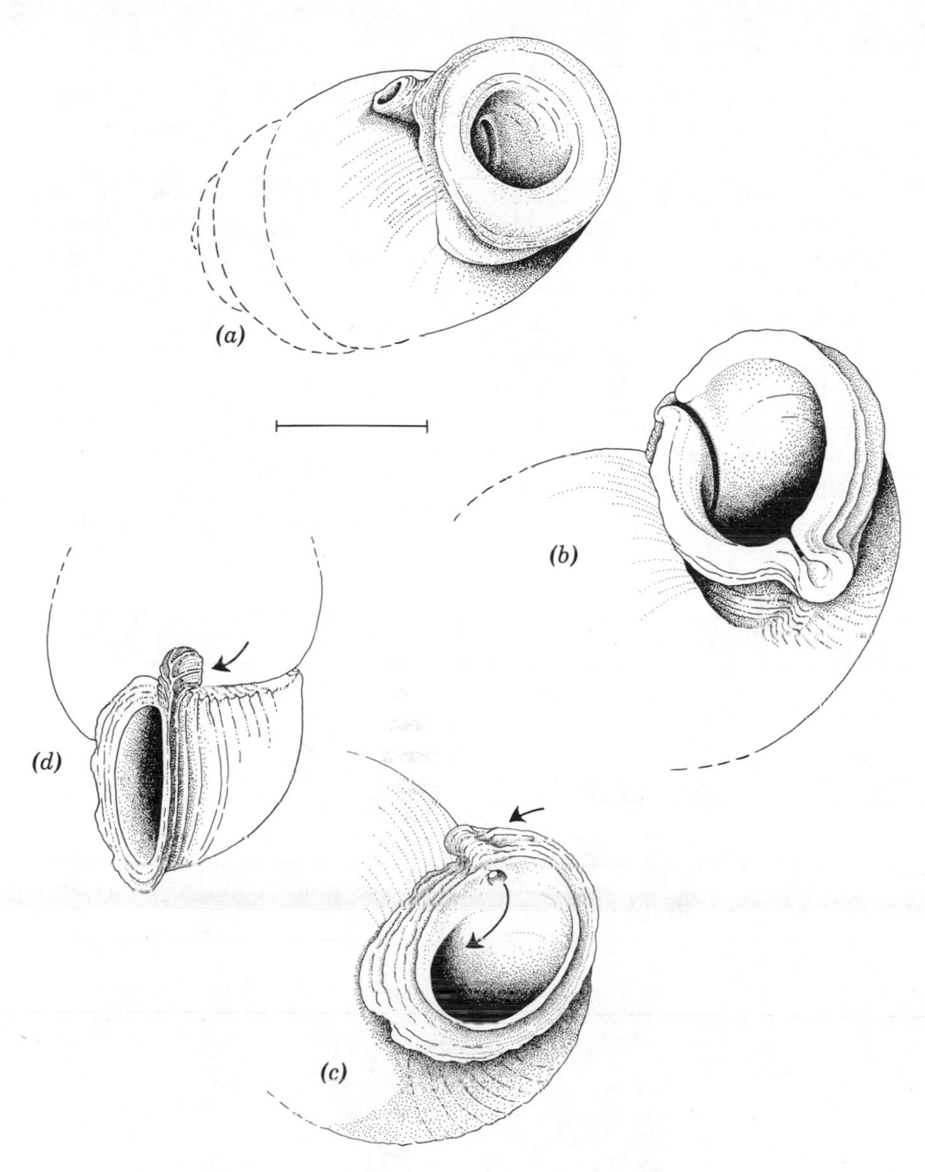

FIGURE 5. Breathing tubes in prosobranch land snails: (*a*) *Rhaphaulus lorraini* from Bukit Besar, Jalor, Malaya; (*b*) *Pupinella simplex* from Mt. Perry, 70 miles west of Bundaberg, Queensland, Australia; (*c, d*) *Opisthosiphon bahamensis* from Breezy Hill, Nassau, New Providence Island, Bahamas. Scale line equals 5 mm for *a*, but only 2 mm for *b* and *c*; d is on a slightly smaller scale than *c*.

ward, while a second notch on the upper margin of the aperture provides a more shallow, supplementary breathing pore if the snail is retracted only slightly.

All of these examples are from the Superfamily Cyclophoracea. They are most closely related to the freshwater families Viviparidae and Ampullariidae. Probably this is the most primitive living group of the mesogastropod mollusks. While specialized in living on land and in freshwater habitats, their anatomy, except for adaptations to land life, is generalized.

Two other prosobranch groups have made the transition to land, and one of these, the West Indian Pomatiasidae, also has species that developed breathing tubes. *Opisthosiphon bahamensis* (Fig. 5*c*), for example, has a breathing tube covered with a reflected hood that presumably prevents dirt particles from sifting into the shell (5*d*).

Pulmonates do not have an operculum to seal the aperture. Instead of an operculum, they secrete a membrane of mucus or calcified mucus across the shell opening which is called an epiphragm. In forms such as *Helix aperta* (Fig. 6, lower row), the thickness and strength of this membrane almost equal those of an operculum. The membrane allows air to filter back

FIGURE 6. Apertural closure in prosobranchs and pulmonates. Upper row: *Cyclophorus* sp., an arboreal prosobranch from Burma, specimen on right has the operculum in place; lower row: *Helix aperta* from France, specimen on right has heavily calcified epiphragm in place. Courtesy Field Museum of Natural History, Chicago.

and forth, but prevents loss of water. Sometimes a pulmonate will secrete a series of up to five or six such barriers, the first just at the shell opening, then subsequent ones at short intervals as the snail retreats deeper and deeper into its shell. The large number is characteristic of desert or monsoon climate pulmonate snails. Those species living in the moist woodlands of Europe or Eastern North America will generally secrete only one or two thin epiphragms.

Both prosobranch operculum and pulmonate epiphragm are means of hiding from a dry spell. But they do not explain the greater success of the pulmonates on land. This results from changes in structure of the pallial cavity. First of these was fusing the mantle collar to the neck of the snail. This left a large internal chamber connected to the outside by the small hole called the pneumostome. Instead of the entire pallial surface being exposed to the air, and losing water by evaporation, air is pumped in and out of this narrow opening. The loss of water is much less under these controlled conditions. In addition, having the pallial cavity enclosed allows it to be put to another use, that of carrying a reserve supply of water. From a drop or two to a quantity equal to one-twelfth the total body weight of the snail can be present within the pallial cavity. This water can be taken into the body if the extended foot and body are losing water from evaporation, or water from the body can be added to the pallial reserve.

The margin provided by carrying a water supply around with it permits the pulmonate snail to be active far more often than the prosobranch. A pulmonate snail can remain active at much lower humidity, continuing to feed when the prosobranch must be inactive. The pulmonate also can store a significant quantity of water in its shell when it does go into aestivation and thus is in better position to wait for the return of favorable moisture. Since it can be active at lower humidity, the pulmonate will return to activity sooner than the prosobranch.

When in hibernation or aestivation, both types of snails react similarly. The heart of the pulmonate *Helix pomatia* may slow from 36 beats per minute to only three or four, while in the prosobranch *Pomatias elegans* it may drop from 53 to only two to three times per minute. As an indication of how much body processes are slowed down, a hibernating *Helix pomatia* at 30°F will use as little as $\frac{1}{50}$th the amount of oxygen used by an active snail at 59°F.

Other changes in the pulmonate snails include the evolution of tentacles that can be withdrawn into the body, and several improvements in the structure of organs in the pallial region. The latter changes permitted slugs to evolve from snails. This story is covered in Chapter XII.

Snails could venture onto land because the shell provided a place for retreat and the pallial cavity contained empty space that permitted the head and foot to be withdrawn into the cavity to minimize loss of water. Prosobranchs have retained the operculum and remain restricted to wetter areas of the world. They can survive periodic droughts by means of the very effective apertural seal provided by the operculum and accessory breathing tubes. Pulmonates have a means of reducing water loss and carry a "spare tank" of water with them by enclosing the pallial cavity. Their greater diversity on land results from water conservation.

XI
Experiments in Living

Colonization of the land opened up a world of opportunities and presented several new problems. All land snails have problems of locomotion, as locomotion is far more difficult on land than when the body is supported mostly by the water. Self-protection was also more difficult. Prosobranchs could retreat behind the operculum and gain protection from enemies, but pulmonates had to find other means of protection against being eaten. In rivers and oceans swirling currents of water wash the shell and only the problems of encrusting organisms exist, but snails crawling through moist leaf litter and humus may have particles of dirt adhere to their shell. The smaller the snail, the greater the problem, since at very small sizes the strength of adhesion will be greater than the pull of gravity. How can accumulations of granules on the shell surface be limited?

Once past these initial problems, there were opportunities. In the moist tropics there are trees with heavy growths of algae and fungi, a wonderful source of food. There are also exposed rock surfaces covered most of the time with seepage water and a tangle of lichens and tiny mosses. Tidal zone snails had long since experimented with internal fertilization and/or keeping the fertilized eggs inside the snail's body until they hatched. Given the great amount of abundant food and steady moisture supply in some areas of the world, the luxury of experimentation was possible. Following the oft-quoted but seldom written law of evolution that "When it comes to sex, to hell with utility!," eyebrow lifting structures and behavior patterns emerged.

Times and climates change. We now live in an era of markedly varied weather conditions. Areas that 10,000 years ago were moist forests are now cactus-studded deserts. The snails in the formerly moist forests have sometimes lingered on, first retreating into the last remaining pockets of moisture, then chance mutations permitting survival. Natural selection led to the loss or size reduction of genital structures, longer and longer periods of aestivation occurred between the infrequent rains, the number of young produced became smaller, and the length of life greatly extended. These are only a few of the ways in which snails have changed and adapted to life on land. They offer a glimpse into diversity and the ways in which a major group of animals undertook experiments in living.

LOCOMOTION

Except for such specialized groups as the conchs with their lurching progression (Fig. 8 in Chapter VI), marine snails move by a combination of cilia beating against a trail of mucus laid down by the snail and waves of muscular contractions. Freshwater pulmonates apparently move solely by ciliary action, but the land snails show a diverse series of locomotor types. A generalized pattern is shown both by the archaeogastropod *Helicina* (Fig. 1*a*) and advanced pulmonates such as *Helix* (Fig. 1*d*). They glide on the sole of their broad foot by means of a series of ripplelike bands that proceed from tail to head. The dark moving bands are areas where the foot is lifted slightly off the surface of the ground and thrust forward fractionally. Internally a highly complicated system of muscles and hydraulic action combines to raise, move forward, and then lower the foot to the slime trail. Many variations of this pattern exist in which the locomotor waves run either front to back or on only part of the foot sole.

Land prosobranchs show two major modifications. In pomatiasids such as the Jamaican *Colobostylus* (Fig. 1*b*), the snail walks only on the edges of the foot, which has an inverted U-shaped cross section. A wave of contraction moves first along one side of the foot from back to front, and then a similar wave starts on the other side. First illustrated by Amos Brown, this has been seen subsequently in most members of the family. The animal "waddles" along, swaying from side to side, with its operculum acting as a saucer-shaped support for the shell spire. Two tiny trails of mucus are left behind and readily "tracked." In contrast, *Truncatella* (Fig. 1*c*) uses

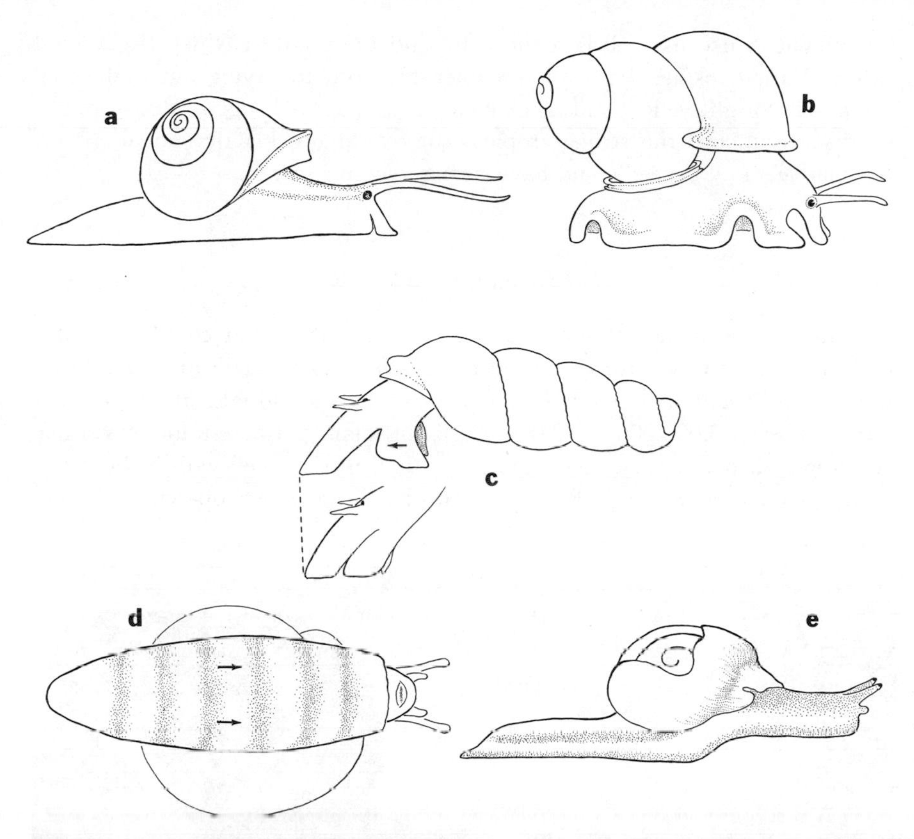

FIGURE 1. Locomotion patterns in land snails: (*a*) *Helicina;* (*b*) *Colobostylus;* (*c*) *Truncatella;* (*d*) *Helix* in bottom view to show locomotor waves; (*e*) *Velifera.* Not drawn to scale. *a–c* after Amos Brown.

a modified conch lurch. Its proboscis is stretched out, the foot and shell pulled up, then the proboscis extended again.

Generally the term "snail's pace" is appropriate, with movement of only a few inches each minute quite common even for a large snail or slug. But small relatives of *Colobostylus* can move seven times their shell length each minute, and a few larger land snails have been reported to "lope" or "gallop" along by creating extraordinarily larger locomotor waves in the foot. The most surprising snail I have seen is a small (shell diameter $\frac{1}{8}$ inch) Costa Rican mountain snail, *Velifera* (Fig. 1*e*), which, when in a hurry, raises the front half of its foot off the ground, gliding only on the posterior half. A specimen traveled an "incredible" 18 inches a minute when exposed to sunlight. Since its shell is reduced beyond the point at which the animal can withdraw inside, the snail is vulnerable both to drying out and being eaten. The bright yellow and black colors of the animal suggest that it might be distasteful, but the strong tropical sun could quickly dry out *Velifera.* Its relatively great speed could have strong selective value.

LIVING IN LITTER

The upper few inches of soil and the decaying litter that covers it swarm with millions of tiny arthropods and nematode worms. Snails are a relatively minor constituent of the soil fauna in terms of individuals, although their numbers reach 3,000,000 to 12,000,000 living snails per acre under favorable conditions. Most of these snails are very small in size, well within the range many predaceous arthropods would consider as food. An operculated snail

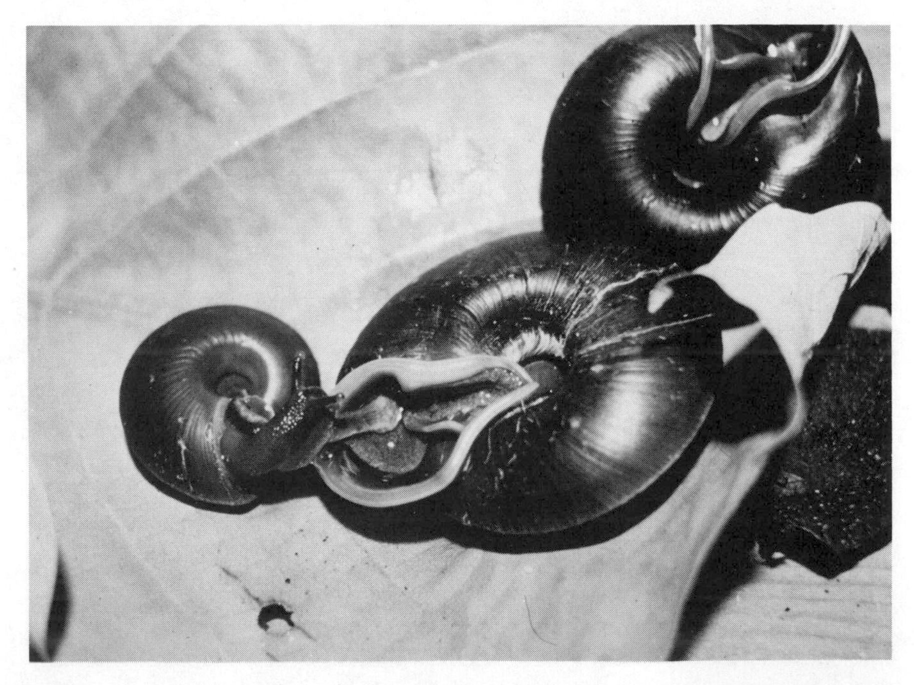

FIGURE 2. Young and retracted adult of the Panamanian camaenid land snail *Labyrinthus otis,* a species that develops huge apertural barriers at maturity. Photograph by the author, courtesy Field Museum of Natural History, Chicago.

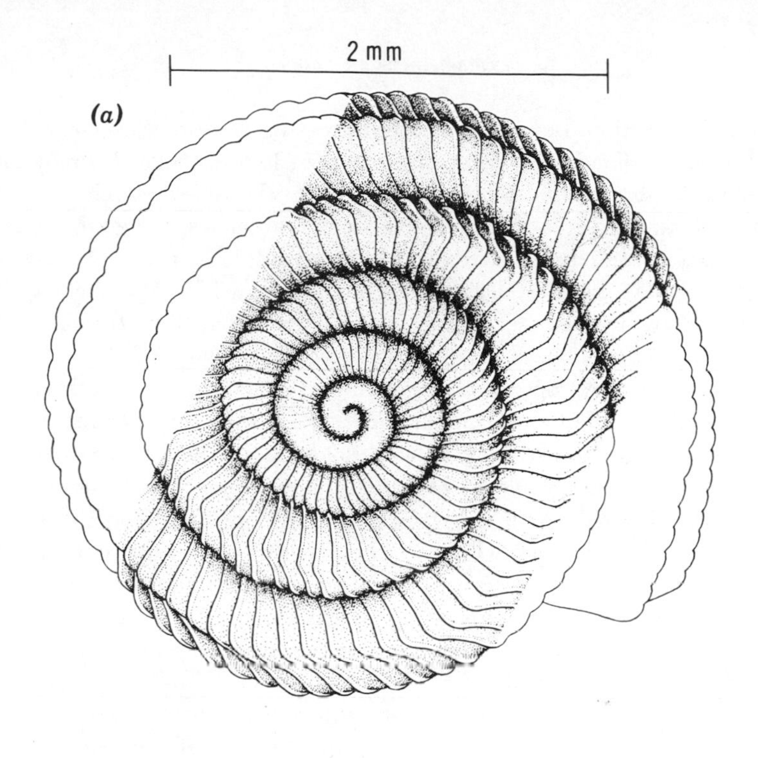

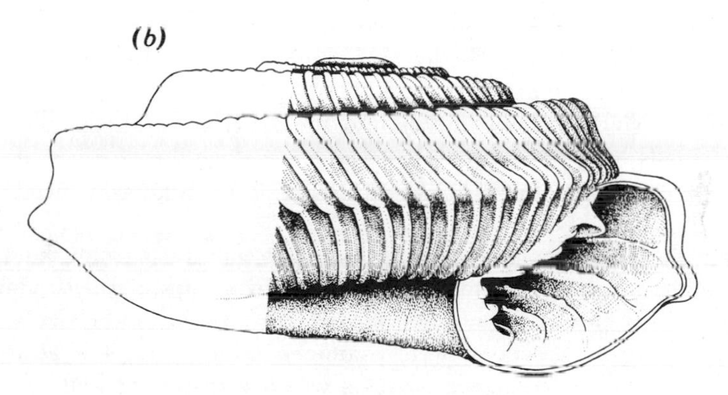

FIGURE 3. An endodontid land snail of the genus *Thaumatodon* from the Lau Archipelago of Fiji: (*a*) top view of shell; (*b*) side view of shell. The apertural barriers and major surface sculpture are shown, but the microsculpture is omitted. Drawing by Carole W. Christman, courtesy Field Museum of Natural History, Chicago.

has some protection, but a pulmonate retreating into its shell could be followed by a small mite or insect and eaten at leisure. Indeed, firefly larvae and many carabid ground beetles prey on snails almost exclusively.

Many of the larger and more advanced pulmonates survive by secreting noxious mucus or climbing tree trunks to avoid little predators, but most families of snails have at least a few species in which the shell aperture is narrowed by a series of *barriers*. These barriers can take the form of tubercles, long ridges, knobs, or indented apertural margins. Their function is to narrow the shell opening. The snail can squeeze in and out, but hopefully the opening will be too narrow for a predator to enter. In many species these barriers are formed only at adulthood, as in the Panamanian *Labyrinthus otis* (Fig. 2), while in others they are present throughout the life of the snail. Typical of the latter are the endodontid land snails of Pacific Islands. In *Thaumatodon* (Fig. 3) the barriers are numerous and high enough so that the aperture is only a fraction as wide as it would be without them. These barriers grow throughout the life of the snail, added to at the anterior end and resorbed posteriorly. In this way they are always located just inside the shell aperture, but do not clutter the interior, leaving room for the retracted snail. If the shell of an endodontid is broken and the surface of these barriers looked at with a scanning electron microscope, many are found to have a sculpture of triangular points on top of the barrier (Fig. 4). These points are aimed toward the outside of the aperture. They would tend to tangle the hairs on the feet or antennae of an arthropod. The predator could free itself quite easily by backing out, but it would get into more trouble by trying to crawl further into the aperture. Perhaps 7000 species of land snails have barriers developed at some stage during their life. They are more commonly found in small species than large ones of the same group.

Although most apertural barriers have a sculpture of some kind, these sharp points are found on the barriers in a few groups of snails. How they originated was puzzling, until the barrier surfaces in larger snails, which have very narrow apertures, were examined. Since a snail is attached to its shell only by the columellar muscle, which is anchored well behind the aperture, the animal can pull itself *into* the shell, but getting back *out* re-

---

FIGURE 4. Barriers from the inner wall of the aperture in an endodontid land snail from Rapa Island, Austral Group, South Pacific: (*a*) top of barrier showing detail of "points" at 1000×, where the background "bumps" represent the primitive "gripping" sculpture; (*b*) posterior margin of two barriers at 325×, showing how the back margin of barriers and callus (lower left) are resorbed. Courtesy Field Museum of Natural History, Chicago.

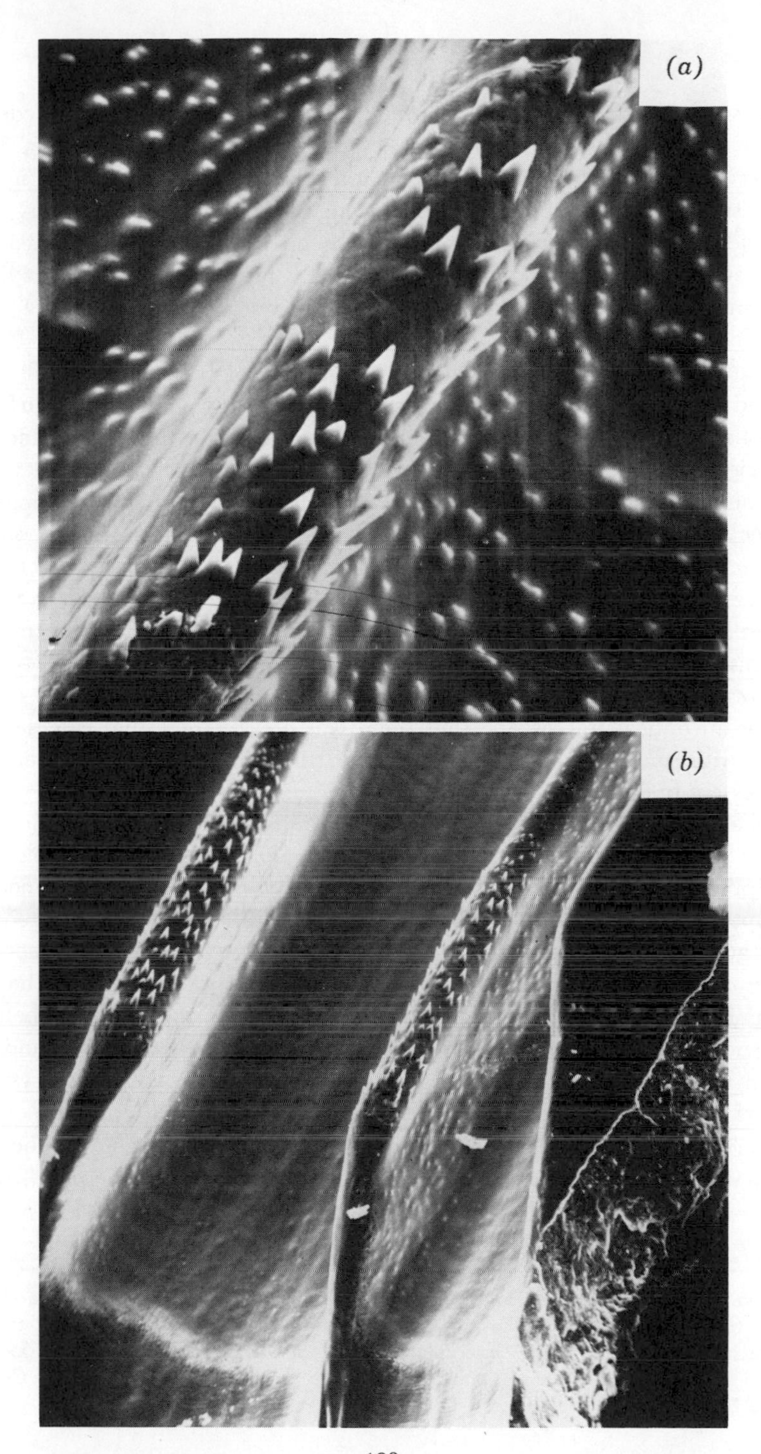
(a)
(b)

quires a different set of muscles. *Stenotrema barbatum* (Fig. 5) is a common $\frac{1}{4}$ inch diameter snail found in wet streamside forests of Eastern North America. Only a very narrow apertural slit remains in adults (Fig. 5*a*). The surface of the lip is "shingled" (Fig. 5*b*) with little platelets. These provide microgripping surfaces that the snail can use to pull its head and foot out of the extremely narrow opening. Evolution of platelets that helped the snail extend its body into points that discourage a predator is a logical step.

Equally striking in appearance is the surface of many litter snails (Figs. 6 and 7). Many insects and limacacean snails use a smooth, waxy surface to repel moist particles, but other groups try to break up the shell surface into a series of fine points. A wet bit of rubble touching a few raised points might not adhere, whereas if the surface was smoother, the two might stick together. Insects and some snails have bristles protruding from the surface. Young of *Stenotrema barbatum* (Fig. 6*a*) have regular rows of bristles projecting far above a low relief series of wrinkles (Fig. 6*b*). The endodontid snails take an entirely different approach. Here sculpture added to sculpture provides defense in depth against particle adherence. Adult New Zealand *Ptychodon microundulata* (Fig. 7) average $\frac{1}{11}$ inch in diameter. It could be seen by using an optical microscope that there was a series of radial ridges on the shell with something in between. Prior to the scanning electron microscope the extent to these fine details was not even imagined.

Shell sculpture varies widely from group to group, but the general pattern holds that the smaller the snail, the more complex and closely spaced is the sculpture. If the adult snail is somewhat above a size of about $\frac{1}{5}$ inch, the sculpture tends to be less complex and reduced in prominence. But reducing the adherence of particles is not the only function of shell surfaces. The litter zone of forests and fields contains decaying organic matter. The weak "humic acids" produced during decay would erode the calcium shell if it did not have a protective surface layer. As shown in a charopid land snail from the Tonga Islands in the South Pacific (Fig. 8*b*), the inner crystalline layers of calcium (lower part of photograph) are covered by an impervious, noncrystalline organic layer called the *periostracum*. Composed of chemicals very similar to those that form the body covering of insects, the periostracum

---

FIGURE 5. Shell aperture and "gripping" sculpture in the polygyrid land snail *Stenotrema barbatum* from Eastern North America: (*a*) aperture showing the very large barriers, at 25×; (*b*) microplatelets on the barrier surfaces, at 2575×. The raised edges of the platelets point toward the outside of the aperture. Courtesy Field Museum of Natural History, Chicago.

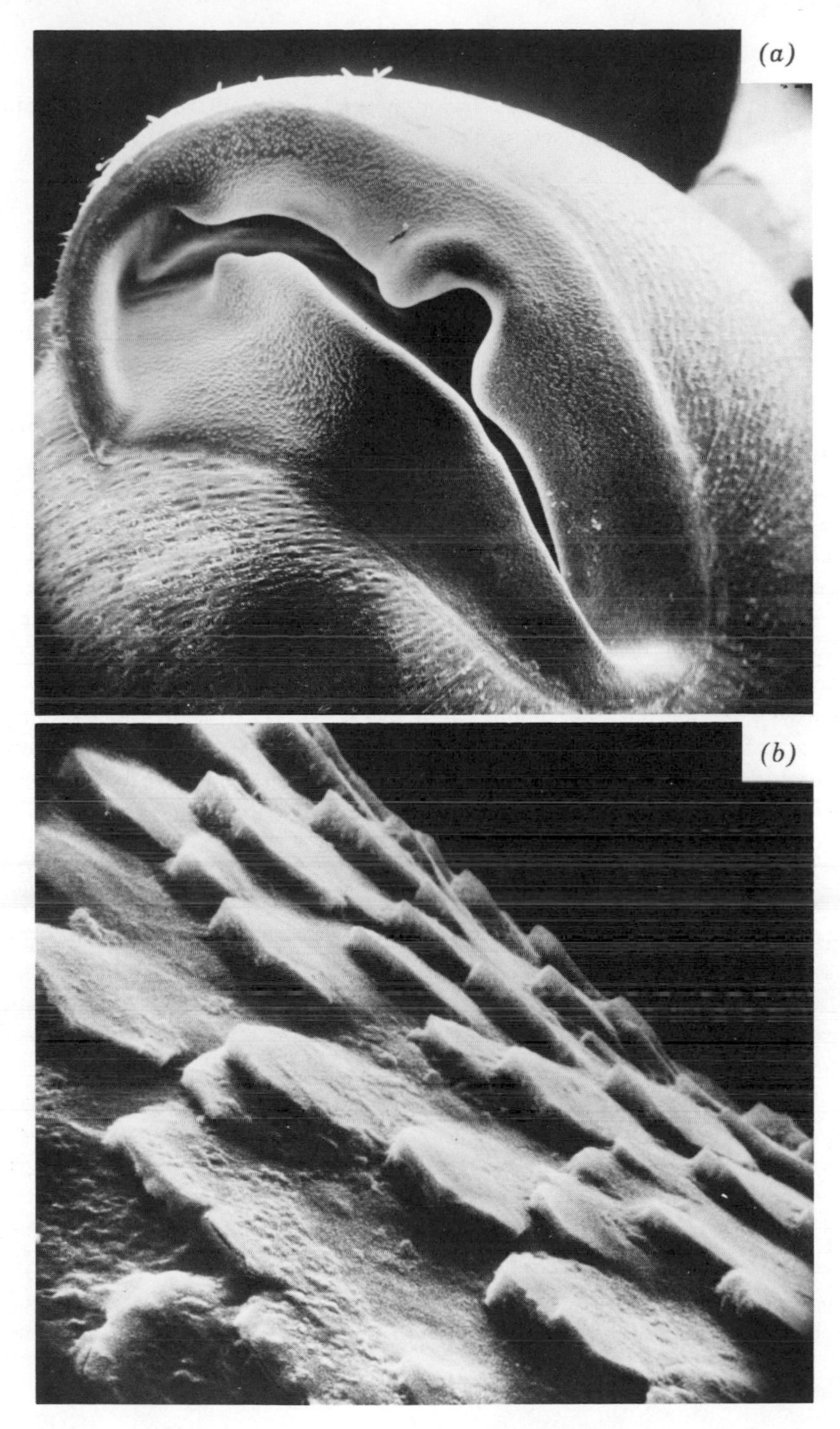
(a)
(b)

protects the calcium layers from erosion by acids in the environment. In many species it is the layer that sculpts the complex shell surface features.

ABOVE THE GROUND

In temperate and tropical forests where moisture is not a problem and the luxuriant growths of lichens, algae, and fungi provide food in abundance, many snails have come up out of the litter. Perhaps an early step was to crawl over moist rock faces. Today in the forests of such diverse areas as Central America, Cuba, Jamaica, Borneo, Tennessee, Malaya, and Central Africa, limestone outcrops over which water drips harbor a fantastic variety of snail life. Many are very small. At first these snails may be sampled by collecting bags of debris from the base of cliffs or the sides of streams and patiently sorting out the micromollusks. A sample of such debris from a limestone hill in Borneo (Fig. 9) yielded literally thousands of snail shells.

In time people began to investigate exactly where these snails live and how they divide up the environment. During a brief visit to Malaya, I collected several species at Bukit Chintamani, a large isolated block of limestone near Kuala Lumpur. The photographs in Fig. 10 are of a rock face covered by a mass of mosses and lichens. Arrows (Fig. 10*a, b*) point out a minute prosobranch, *Opisthostoma retrovertens,* in the middle of the moss, and a slightly larger pulmonate, *Gyliotrachela depressispira* (Fig. 10*b, c*), which occurs in more open lichen-encrusted spaces on the rock surface. The two genera maintain this pattern of microhabitat differences wherever they occur. Why they develop the peculiar growth pattern (Fig. 10*d, e*) on the last whorl is unknown. Conceivably the resulting shape might reduce the chance of being swept off by a surge of water when fastened to the rock surface. All specimens of *Gyliotrachela* I saw were oriented in the same position, with the aperture above the spire, so that a curved surface would meet descending water.

Other species will roam in search of food at night, but return to very restricted microhabitats to hide during the day. Carnivores such as *Discartemon* (Fig. 10*f*) and *Oophana* fall into this category. For example, one species of *Oophana* has been found only on Bukit Serdam, near Raub, Malaya. Dead shells had been collected for years, but my wife achieved

---

FIGURE 6. Shell sculpture on a young *Stenotrema barbatum,* showing the nature and extent of periostracal processes: (*a*) pattern of "hairs," at 195×; (*b*) detail of a single hair and the surface ridges, at 360×. Courtesy Field Museum of Natural History, Chicago.

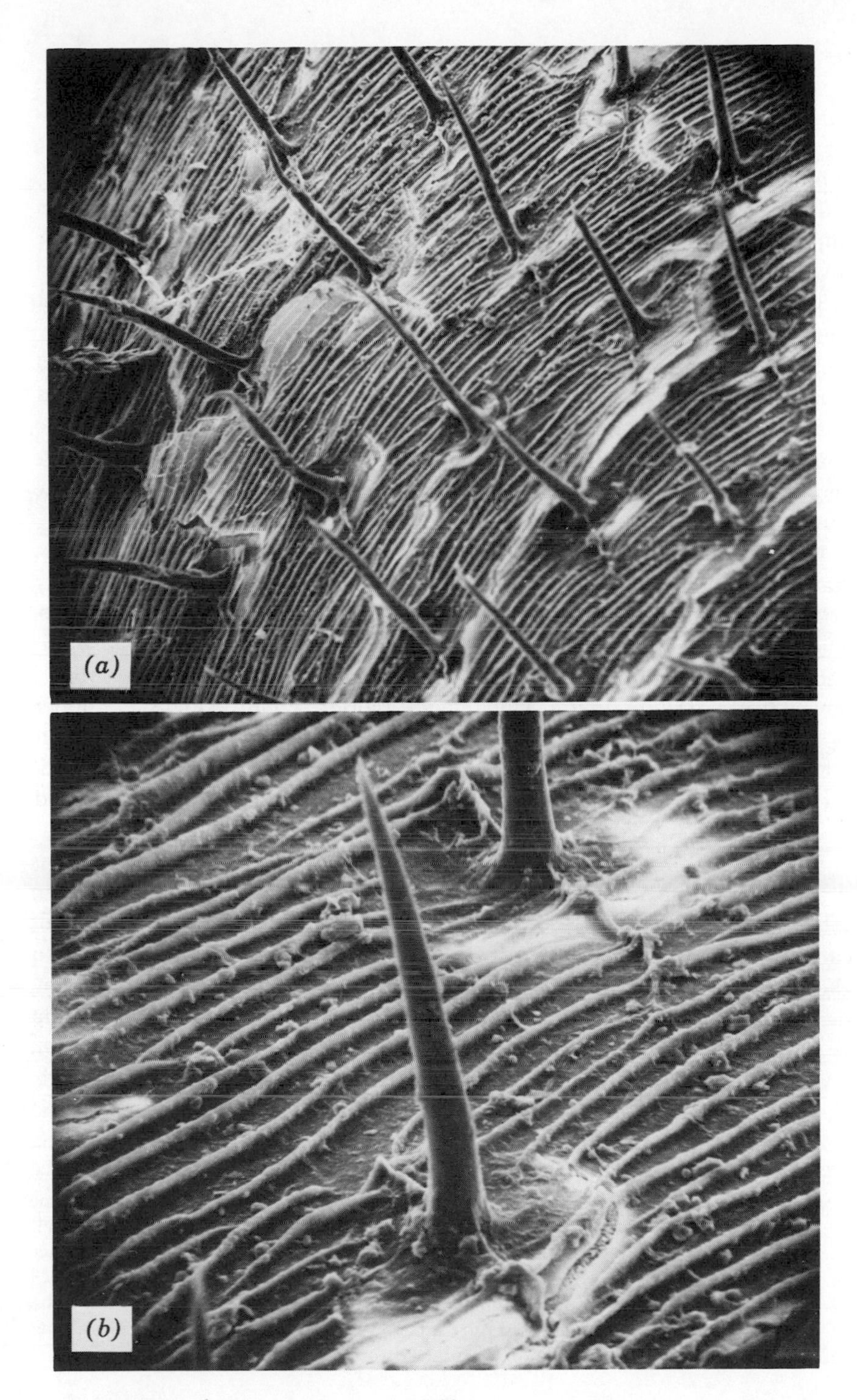
(a)
(b)

a malacological first by discovering that live specimens could be found during the day in small V-shaped leaf-mold filled crevices on top of fallen limestone rocks, while myself and another trained malacologist were hunting with total lack of success in more "traditional" niches a few feet away. This emphasizes both the diversity of snail habitats and hiding places plus the advantages of a fresh outlook to a problem!

Snail exploration above ground level did not stop with rock faces. In wet seasons tree trunk dwellers such as *Trochomorpha* (Plate 12, third from top) are found out and crawling, with the species in Europe and Southeast Asia being particularly obvious and numerous. However, where a cold season occurs, true tree dwelling mollusks could not evolve. Their bodies are too small and their temperature-regulating ability is virtually nonexistent. A sudden early fall frost may catch Middle Western land snails partway up the trunks of trees and kill them by freezing.

It is only in the true tropical lowland areas, where frost never comes, that fully arboreal snails evolved and flourished. Large in size and usually brightly colored, they are among the most beautiful of all molluscan shells. They feed mainly on algae and lichens, are active during damp weather, and seal to a tree limb or leaf during dry spells. Some descend to the ground to lay their eggs, as do the Cuban and Florida *Liguus* (Plate 11, upper right), while others roll a leaf into a basket to form a cradle for the eggs (the Indonesian *Amphidromus*), or retain the eggs inside the body until they hatch as crawling young (the Hawaiian *Achatinella,* Plate 12, second from top and lower right).

For many years the purpose of their bright shell colors was debated. Was it simply a means of getting rid of body waste products by converting them into pigments and depositing them in either the periostracal or outer calcareous layers of the shell? Was it a confusing coloration pattern, since often there were stripes or color bands developed? Or was it a warning coloration, designed to tell a predator that this was not a tasty morsel? The Florida *Liguus* is known to secrete liquid that drives off ants, but nothing is known concerning the other species. Unfortunately it may not be possible to study these much longer, since the clearing of forests in Florida and Indonesia

---

FIGURE 7. Surface sculpture on a charopid land snail from near Wellington, New Zealand: (*a*) view of shell surface at what would be maximum dissecting microscope observation (102×), the juvenile shell being only 1.38 mm in diameter; (*b*) detail in area of postnuclear growth seen in lower left of *a,* at 310×; (*c*) shell surface on outer edge of shell (upper right of *a*) at 1050× showing four of the 124 radial ribs present on the body whorl; (*d*) sculpture between two of these ribs, at 3215×; (*e*) detail of microsculpture at 13300× showing an inner even finer sculpture. Courtesy Field Museum of Natural History, Chicago.

(a)

(b)

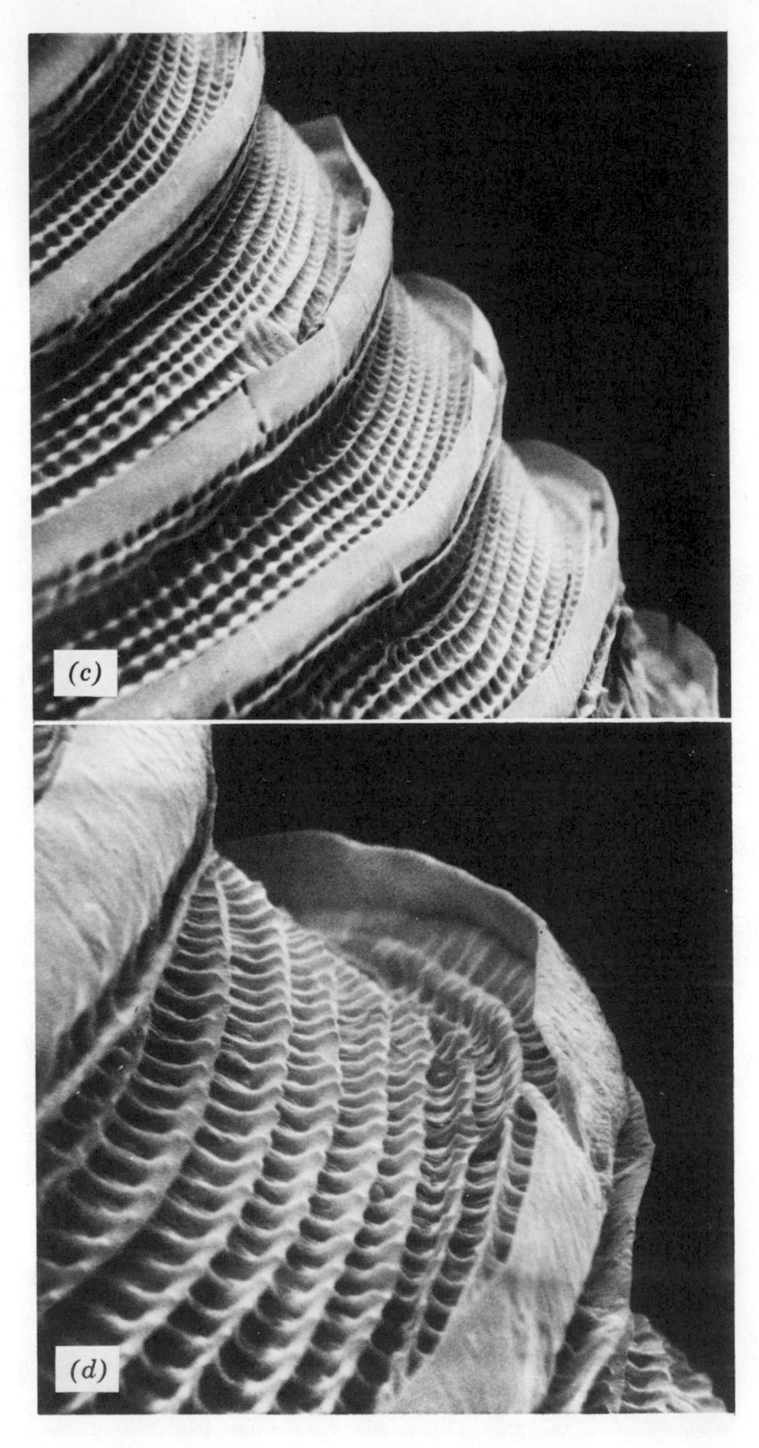

(c)
(d)

(e)

may soon cause many races of *Liguus* and *Amphidromus* to become extinct. The Hawaiian *Achatinella* used to be made into leis (necklaces) but with the destruction of native forests and ravages of introduced snails, they are now extremely scarce. Several species are already extinct and others are endangered and on the verge of extinction. Despite their similar shapes, the tree snails from different areas of the world are not related to each other. The Bulimulidae in Florida, Melanesia, and South America, the Camaenidae in Indonesia and Melanesia, the Bradybaenidae (*Polymita,* Plate 11, and *Helicostyla*) in Cuba and the Philippines, some Trochomorphidae and Partulidae on Pacific Islands, and the Achatinellidae and the usually smaller but equally colorful Amastridae in Hawaii all represent independent evolution into tree snails. They differ in feeding, genitalia, life history, and activity patterns, sharing only bright color and usually an elongated shell shape.

SEX

A prerequisite for land life independent of a return to water for reproducing is internal fertilization coupled with retention of the eggs until they are ready to hatch (ovoviviparity), giving birth to live young (viviparity), or laying eggs with a hard covering (oviparity). Among the oviparous species are the South American land snails of the Genus *Strophocheilus,* which reach a length of 6 to 9 inches and lay oval hard-shelled eggs up to $1\frac{1}{2}$ inches in length. Many snails are ovoviviparous (Partulidae and Achatinellidae) and a very few are viviparous.

The preliminary to successful egg laying is successful mating. Vertebrates depend on visual or sound signals for individuals to recognize potential mating partners, and, once recognized, they use stereotyped behavior patterns to woo a mate. Many insects depend on chemical signals followed by "courtship flights." Snails do not see very well and have no ears, so they depend on a sense of touch coupled with behavior patterns to find a suitable mate. If several closely related species live in the same area, species differences in structures and behavior are needed in order to separate like and nonlike partners. Hence elaborate supplementary genital structures have evolved.

FIGURE 8. Shell sculpture in a charopid land snail from Tonga: (*a*) pattern of surface sculpture, at 640×; (*b*) cross-sectional view of a single large rib at 1150× showing the inner (lower) calcareous layers that make up most of the shell thickness and the noncrystalline periostracum or outer (upper) layer that both protects the calcium from acids in the leaf litter and forms the complex surface sculpture on the shell. Courtesy Field Museum of Natural History, Chicago.

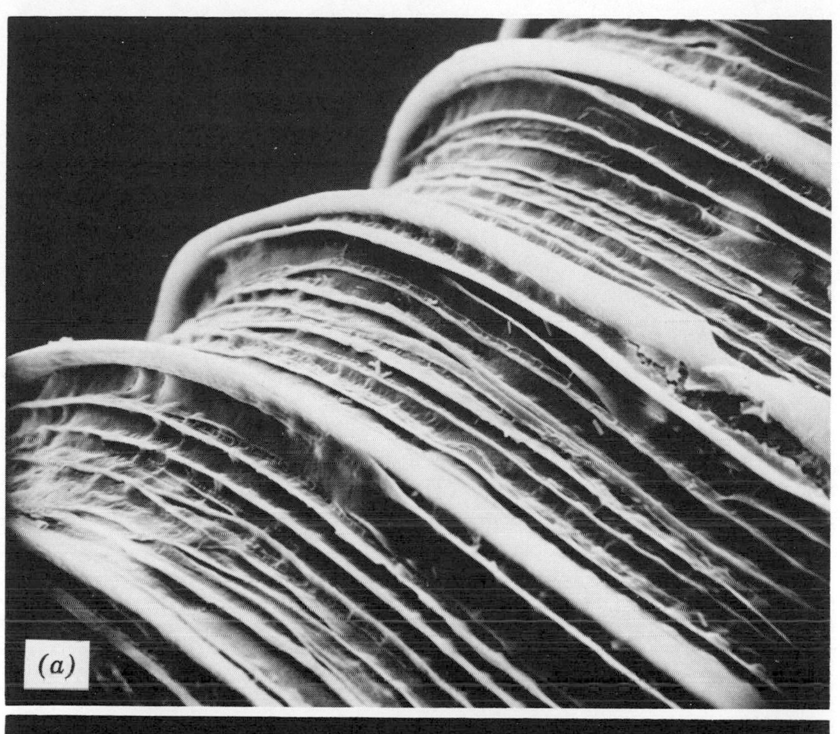
(a)

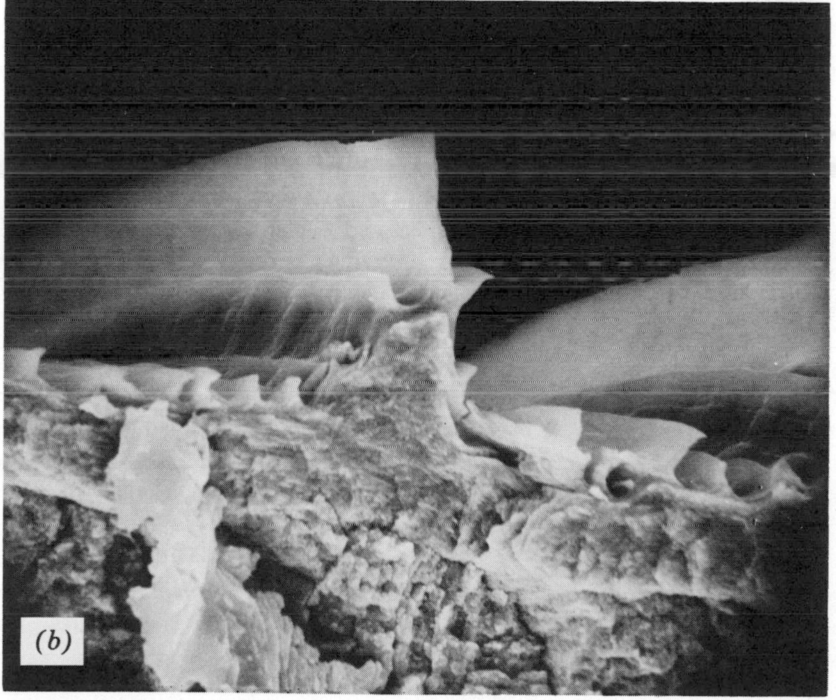
(b)

FIGURE 9. Land snails from a Bornean limestone hill. The watch crystal is 2 inches in diameter. Courtesy Field Museum of Natural History, Chicago.

A very frequent pattern is to develop a *dart apparatus*. This accessory structure can secrete calcareous or chitinous spearlike darts that are jabbed into the body of other snails as an "identification" signal. A snail with a 1 inch shell may have a $\frac{1}{3}$ inch long dart. Each species has a dart of a different shape or size so that like can recognize like. Other species depend on soft genital structures being different, while some use a distinctive "sperm packet" (Fig. 11) for species recognition. In many snails sperm will be transferred at one season in such a packet, then stored by the recipient until a later wet spell. Only at this time are eggs fertilized and laid. Sometimes only one mating and one sperm packet are involved, but in some Indian and Southeast Asian limacaceans, up to 20 sperm packets may be found in a single individual. Since a snail can form only a single sperm packet at a time, this means there have been multiple matings with many individuals to accumulate so many packets. The number and spacing of the spines on the sperm packet or *spermatophore* are different for each species living in the same area.

Finally, I wish to report on *Limax cinereoniger,* a common European slug. Most limacid slugs mate while suspended by a string of mucus from a tree limb. They are about 5 to 8 inches in length. When mating, they extend the terminal genitalia from their body. In most species the length of the extruded genitalia will be about one-third to one-half the length of the body. *Limax cinereoniger* normally differs in that the length of the genitalia is almost equal to the body length. In the 1920s a Swiss zoologist, B. Peyer, discovered a variation of this slug on Mt. San Giorgio in the Canton of Tessin, southern Switzerland. The same variant also occurs in moist areas of northern Italy. The genitalia are far longer than usual, the most extreme example of a mating pair (Fig. 12) showing a genital length of 32½ inches from a 6 inch body. As with most land snails, mating is a protracted process. Peyer timed one pair of slugs as spending 19 hours and 20 minutes united together. This is not unusual, since 24 hour matings by many land snails are routine.

The perhaps bizaare sounding extremes in mating just described simply reflect the ultimate result of experimentation in species recognition devices under conditions that are highly favorable for the animal's existence. The Irish elk grew oversized horns, the argus pheasant fantastic tail feathers, and snails did something else.

SNAILS IN DESERTS

When faced with dryness, snails first will crawl toward remaining moisture, then eventually retract into their shells when no moisture differential remains toward which to crawl. They wait for the return of moisture. Either it comes or they die. If 50 out of every 100 die, the population will continue without much difficulty. Even if only two of every 100 snails survive a dry spell, the population can continue. But if only two survive, they will be those whose special genetic makeup enabled them to remain alive longer.

When climates begin to change and a moist area gradually turns to desert-like conditions, many hundreds of generations of snails will be subject to this strong selection. Those snails will be favored who can aestivate longest. Even an aestivating snail requires energy to keep its body tissues in good condition. But if it can get along without parts of some structures, it can save a great deal of energy both in building and maintaining these structures. This can make the difference between survival and drying out. Similarly, producing only one or two eggs takes far less energy than producing 30 eggs of the same size.

In the arid deserts of Southwestern North America are many snails be-

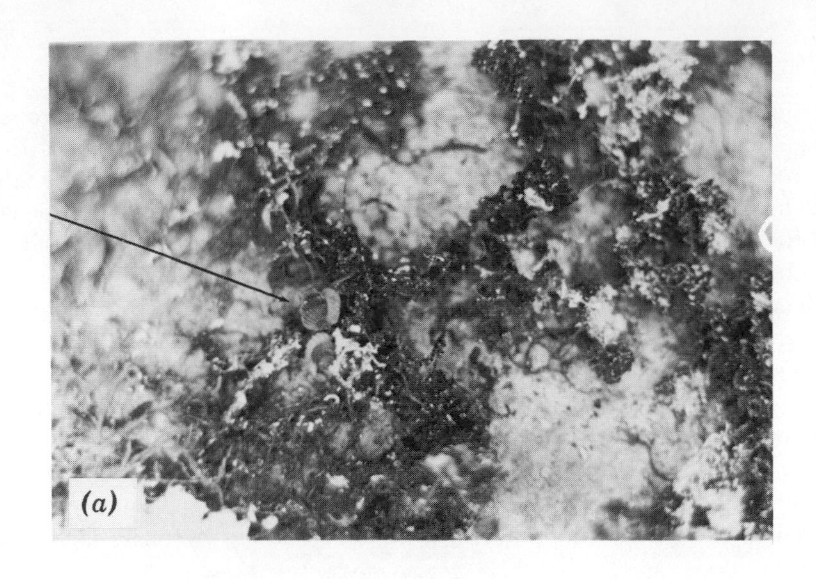

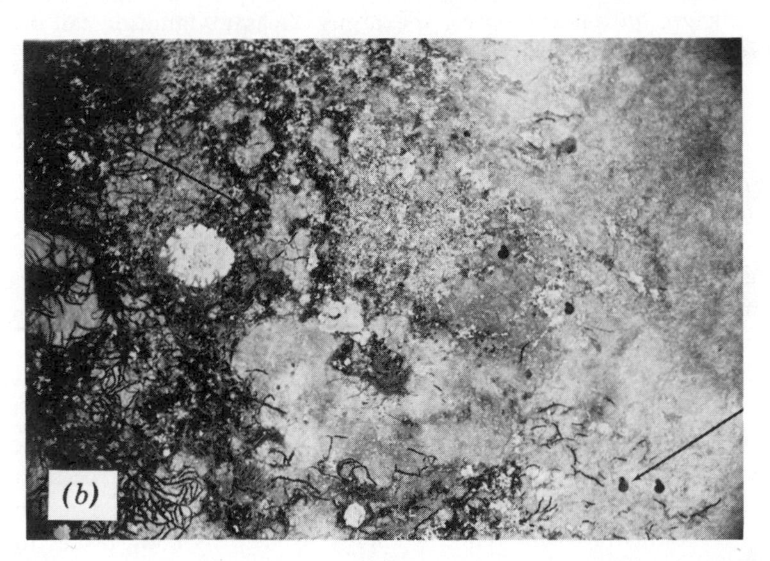

FIGURE 10. Limestone hill snails from Malaya: (*a*) *Opisthostoma retrovertens* (arrow) in moss at twice life size; (*b*) *Gyliotrachela depressispira* (right arrow) and *Opisthostoma* (left arrow) as they would appear viewed from 18 inches away; (*c*) *Gyliotrachela* at twice life size; (*d*) *Opisthostoma coronatum* from Kota Tongkat, Pahang, Malaya; (*e*) *Gyliotrachela depressispira* from Bukit Chintamani, Pahang,

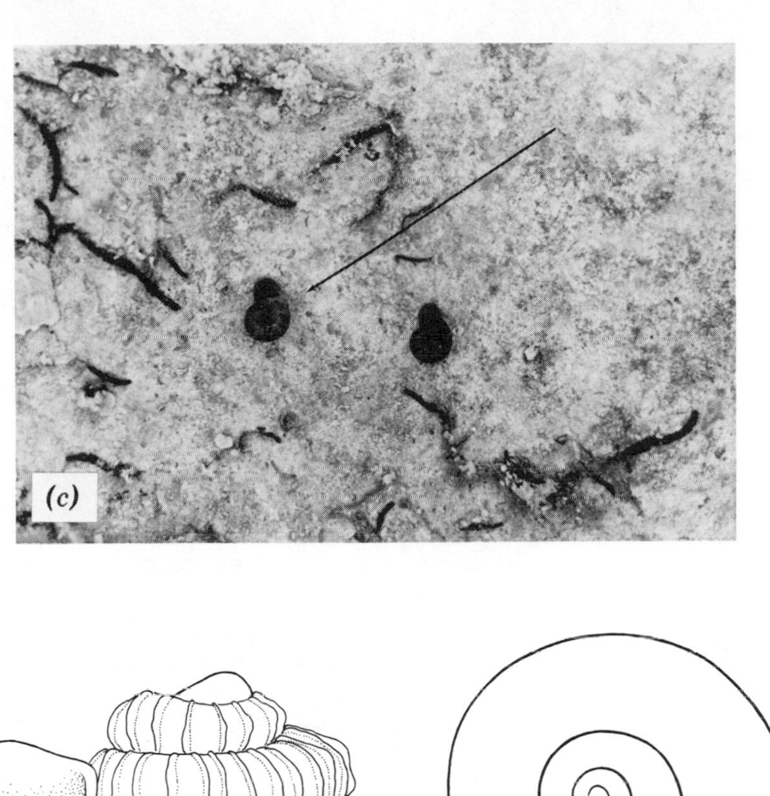

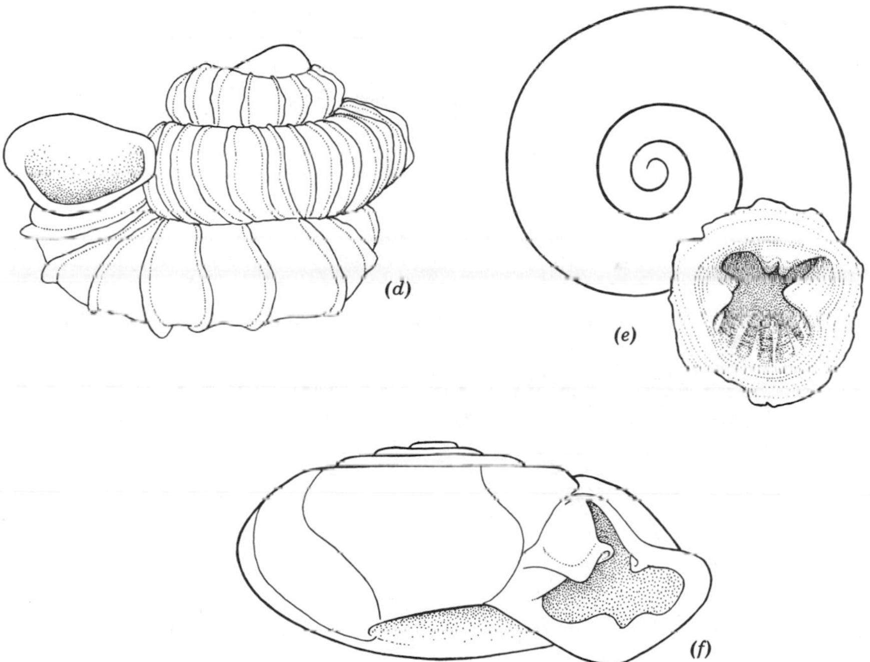

Malaya; (*f*) *Discartemon hypocrites* from Bukit Chuping, Perlis, Malaya. *d–f* adapted from van Benthem Jutting.

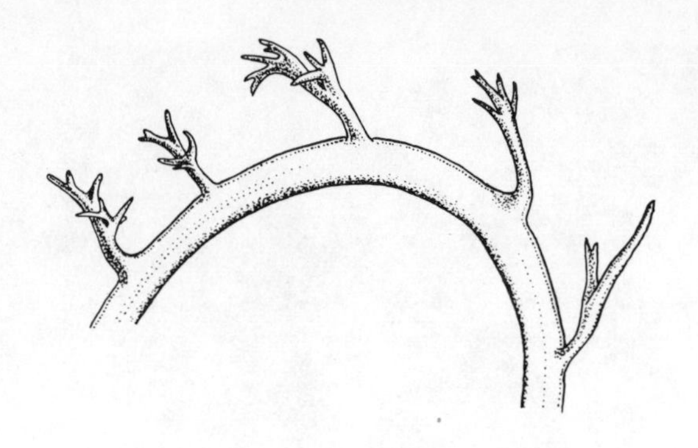

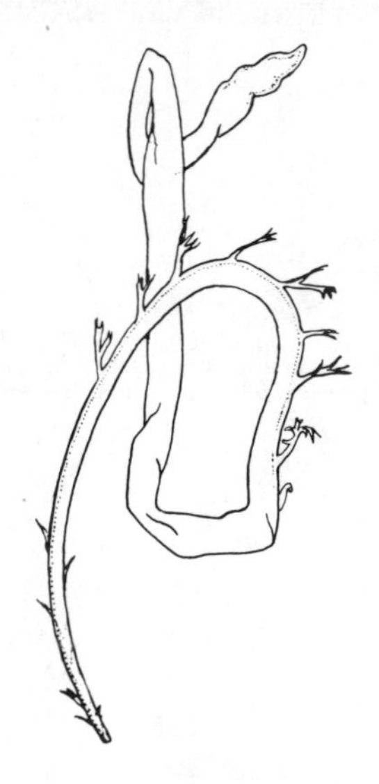

FIGURE 11. Spermatophore from Afghanistan species of *Parvatella*, actual length ½ inch. Detail drawing above shows pattern of spine branching. Each species shows a different branching pattern.

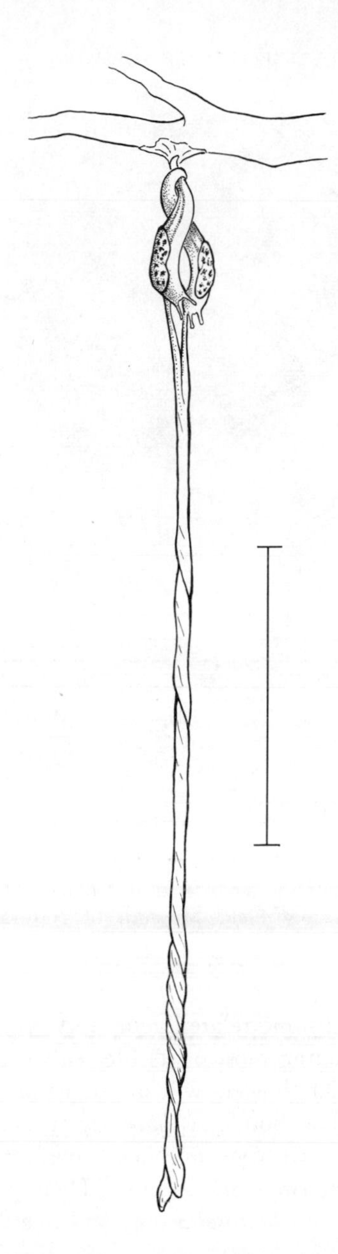

FIGURE 12. Mating slugs, *Limax cinereoniger,* from Mt. San Giorgio, Switzerland. The penis of each specimen is extended, with sperm packets being exchanged at the tips. Scale line equals 12 inches. Adapted from B. Peyer.

FIGURE 13. Live *Sonorella dalli* from the Huachuca Mountains of Arizona. The white "rings" indicate where the snails had sealed onto the rock previously. Photograph by the author, courtesy Field Museum of Natural History, Chicago.

longing to *Sonorella* and some related genera. A typical *Sonorella* (Fig. 13) lives in rock slides, spending most of its life sealed to a rock. Perhaps three or four times a year brief showers will moisten the air so that it can crawl about and feed for a few hours. Apparently it usually will return to the same rock and seal itself with yet another "ring" of mucus. Over the centuries thousands of such rings are formed. Their presence on rocks is the first clue to a collector that *Sonorella* may still live deep down in that rock pile.

Species from drier areas in many cases will have proportionately smaller genitalia, while in the exceedingly dry mountains of inland Southern California there are a series of species that show progressively greater loss of accessory genital structures. This is the reverse of the sexual experimentation outlined above, but has been effective in allowing survival of snails in these deserts. Perhaps most land snails have an annual or two year life cycle, but the West American *Sonorella* can take $2\frac{1}{2}$ to over 7 years just to reach maturity. The polygyrid *Ashmunella kochi ambyla* from West Texas has been kept alive eight years in captivity, and a single specimen of a New Mexican *Ashmunella* lived 14 years after being collected as an adult animal. The initial group of these desert dwelling *Ashmunella* laid only one or two eggs each in their first six years of captivity. This is far below the 100 or more eggs laid each year by many land snails. Such reduced reproductive activity is yet another way in which snails seem to have adapted to the harsh conditions of desert life.

In addition to these rather dramatic and more obvious changes—reduction in genital structures, retreat into moist niches, extended aestivation, and reduced number of young—there are numerous physiological adjustments, changes in shell thickness, frequently a white shell color for those snails remaining in exposed surface areas, and behavioral alterations. As a result of convergent evolution, desert snails from different parts of the world frequently look very much alike, despite having evolved from quite different ancestors.

CARNIVOROUS SNAILS

Many slugs and snails will indulge in cannibalism under crowded conditions or when fed on an improper diet. There are also numerous anecdotal accounts of snails and slugs feeding on dead animals. A few species are true omnivores, eating plants, fungi, and dead animals according to local availability. Some workers in the early 1900s compiled lists of up to 200 plants that a single species of European snail or slug would eat. Apparently some normally herbivorous snails will become carnivorous by preference if given an extended opportunity to feed on meat rather than plant materials.

With this diversity and flexibility in feeding among land snails it is not surprising that some evolved into true carnivores. These species hunt, capture, and kill their prey, which varies from earthworms to other snails and slugs. Hugh Watson demonstrated many years ago that carnivorous snails did not evolve just once, but several times from different lineages. Most snail carnivores share certain features in common: comparatively few, but very long and slender radular teeth (see Chapter IX, Fig. 10), a somewhat enlarged and concentrated nervous system, development of lobes either on the tentacles or near the mouth that act as chemoreceptors, often a slender and quite elongated body that can be stretched far in front of the shell edge (compare the relative body length extension for the carnivorous

Paryphanta busbyi and herbivorous *Achatinella juddii* in Plate 12), and an expanded foot gland. The added sensory equipment, enlarged nervous centers to interpret the data, and bigger foot gland are changes that would aid prey location and swift pursuit. The few large teeth and slender body parallel trends seen in the carnivorous prosobranch snails.

In addition to the actual elongation of the body, there is often a shift in the position of the shell toward the posterior end of the body. This enables the snail to reach further into the apertures of its victims and/or deeper into a narrow crevice or worm burrow. There is not only positional shell change, but often shape and structure change. *Paryphanta,* for example, has the calcareous layers of the shell greatly reduced and the periostracal layers much thickened. Some species of *Paryphanta* have the calcareous layers so very thin that museum specimens are almost impossible to keep whole in a continental climate. Changes in temperature and humidity cause differential expansion and/or shrinkage in the shell layers. This can cause a shell to crack in places, or, indeed, even literally explode into fragments under the stress. I learned first hand of this well-known problem several years ago when a number of *Paryphanta* shells were stored temporarily in my office. A month or so after their arrival I heard what sounded like a BB-gun shot. It took several of these sounds over a month's interval before I discovered that this was caused by *Paryphanta* shells "exploding." During the day steam heat brought high temperature and very low humidity to my office, while at night the heat was off, temperatures sank, and humidity rose.

Many carnivorous snails have a very elongated shell, while in a few genera, such as the Malayan streptaxid *Oophana,* the character of shell growth changes after the first few whorls. The columellar axis of the shell curves so that subsequent growth produces a spectacularly lopsided looking shell. The apex is displaced posteriorly, while the lower whorls lie flatly along the body axis. This lowers the total shell height and permits crawling into narrower crevices.

In addition to narrowing the body, some Rhytididae, such as the New Caledonian *Ptychorhytida* and probably most Australian genera, have the external genital opening shifted from its typical position near the right tentacle and moved halfway or more along the side of the body toward the opening to the pallial cavity. This removal of the several terminal genitalia tubes away from the head and buccal mass permits a substantially narrower body.

Little has been recorded about the actual feeding in most carnivorous taxa. The European slug *Testacella* (see Chapter XII, Fig. 8*g*) will follow earthworms into their burrows. Its radula and buccal mass can be completely everted from the mouth. The long radular teeth then stab or harpoon the victim, which may be pulled whole into the digestive tract as the buccal mass is retracted. The New Zealand *Paryphanta* (Plate 12) uses a different feeding strategy. It will crawl on top of a worm, envelop it in folds of its broad foot, and retract into its shell, pulling the worm inside too. After the worm has smothered to death the snail extends itself, lets the dead worm fall off, and eats its dinner. Other rhytidids, some of the streptaxids, and many of the new world Oleacinidae gradually eat their way up the shell of a victim, at least until the columellar muscle has been severed and then the remaining soft parts can be pulled out of the shell. The victim is gradually consumed. *Euglandina,* a New World oleacinid found from Florida and Texas to Bolivia, sometimes takes 8 to 12 hours before it has munched its way to the apex of the victim.

Among the other land snail families that are exclusively carnivorous are two obscure slug groups, the South African Aperidae and the Indonesia-New Guinea Rathouisiidae, and the North and South American Haplotrematidae. This is not a complete list by any means since various other New World and some Eurasian snail taxa have mostly or all carnivorous members.

This chapter has given only a few of the ways in which land snails have diversified. Equivalent stories can be told about relations between predators and prey in regard to shell color and habitat, the role of snails in soil formation, the problems of flood plain life, and countless other aspects of life on land.

But the selection of topics discussed herein does serve as background for describing one of the most dramatic changes that occurred, the evolution of numerous groups of snails into slugs and semislugs, animals in which the shell either is completely lost or reduced to a fragment buried in the mantle (slugs), or where the shell is a still visible spiral fragment that is far too small for the snail to use as a retreat (semislugs). The problems solved in this evolutionary process are reviewed in the next chapter.

XII
On Becoming Sluggish

Slugs are any snail in which the shell is completely lost or reduced to a fragment that is buried in the mantle. The sea slugs, members of the opisthobranch Order Nudibranchia, were mentioned in Chapter VI. In this chapter are introduced the 500 species of terrestrial slugs and the approximately 1000 species of land dwelling "semislugs."

The idea of a land slug seems a contradiction in terms. Snails survive on land by retreating inside their shell during periods of dryness. Some snails can survive months, if not years, of drought by aestivating inside their shells. The shell serves two purposes; it helps protect the snail against drying out, and also provides some defense against predation by enemies. Yet slugs have no shell at all or only a small fragment buried inside their body. At least a thousand other species of land snails are in the process of losing their shells and "becoming sluggish." Obviously this is a popular trend among land snails. Where this is happening there must be some powerful selective forces favoring any reduction in shell size. Accumulation of many such small mutations over thousands of years eventually would result in a slug evolving from a snail.

For more than 100 years it has been known that all land slugs are not closely related to each other. Slugs have evolved independently from several different snail groups. Scientists still argue about which snails were the ancestors of which slugs. In the process of becoming a slug, the basic anatomy became so twisted and compacted that it is very hard to compare the anatomy of a slug with structures found in possible shell-bearing ancestors. Fortunately there are the many species that are part way between being a normal snail and a slug. Their shell is reduced to the point where the animal can no longer pull its head and foot completely into the shell, but there is still a small shell covering a reduced visceral hump. By studying the anatomy of these "semislugs" it has been possible to discover some of the basic anatomical changes that are a necessary part of "becoming sluggish" and to identify other structures that are relatively unaffected by the change. The latter features can then be used to relate true slugs and shelled snails with fair confidence that actual relationships are being identified.

Just as the process of evolving into a slug has been complex, so introducing this process requires answering several seemingly unconnected questions. What major groups of snails include slugs and semislugs? In what parts of the world did this evolution occur? What environmental factors do these areas have in common that led to slug evolution? What are the basic anatomical features of a land snail? Do the groups that include slugs have special anatomical features? What anatomical structures always change in the course of slug evolution?

Identifying the groups and areas involved enabled recognition of common environmental features of the areas and a few key snail structures that were preadaptations to the evolution of slugs. A presentation of basic land snail anatomy sets the stage for discussing patterns of structure change involved in slug evolution. Finally, it is then possible to survey the diversity of living slugs.

WHO BECAME SLUGGISH

The groups of snails that became fully "sluggish" are not randomly distributed among the land snails, but are concentrated in a few groups. Not including the prosobranchs, there are 60 families of land snails divided into six orders. Two of these orders, the Onchidiacea and Soleolifera, contain nothing but slugs. These animals show many peculiarities of structure and are of uncertain relationship to the other land snails. The basommatophoran family Ellobiidae contains intertidal and a few terrestrial shell-bearing genera. Neither the Orthurethra, which has nine families, nor the Mesurethra, which has four families, have any slug or semislug members. Only in the Sigmurethra can there be found a mixture of shelled, semislug, and sluglike species. Even here the distribution is not random. Only one of the 10 holopod families has any semislugs. Of the 14 family holopodopid group, only the Bulimulidae and Rhytididae have semislugs, while in South Africa the slug family Aperidae evolved from the Rhytididae.

It is only in the aulacopod land snails that there is a great diversity of shelled snails, slugs, and a wild variety of intermediates. Five of the 18 aulacopod families contain only slugs; another family, Arionidae, contains only slugs and semislugs; while three families, Charopidae, Succineidae, and Zonitidae, have mainly shelled and a few semislug species. Two families, Helicarionidae and Urocyclidae, encompass all three types in great variety. These data are summarized in Table IV.

TABLE IV. THE SNAILS THAT BECAME SLUGGISH

Land Snail Group	Shelled Families	Semislugs	Slugs
Superorder Systellommatophora			
Order Onchidiacea	None	None	Onchidiidae (intertidal)
Order Soleolifera	None	None	Veronicellidae (herbivores)
			Rathouisiidae (carnivores)
Superorder Basommatophora			
Order Archaeopulmonata	All	None	None
Superorder Stylommatophora			
Order Orthurethra	9	None	None
Order Mesurethra	4	None	None
Order Sigmurethra		Rhytididae	Aperidae
Suborder Holopodopes	14	Bulimulidae	
Suborder Aulacopoda	18	Charopidae	
		Arionidae	Arionidae
			Philomycidae
		Succineidae	
			Athoracophoridae
		Helicarionidae	Helicarionidae
		Urocyclidae	Urocyclidae
		Aillyidae	
		Zonitidae	Parmacellidae
			Limacidae
			Testacellidae
Suborder Holopoda	10	Helminthoglyptidae	

WHERE DID THIS HAPPEN

Not only are slugs restricted to a few groups of snails but they are also common only in certain areas of the world. The Himalayan slopes of India, Burma, Thailand, and Southwestern China, the high wet mountains of Central and Western Africa, moist regions in the Alps and Caucasus Mountains of Europe, the wet coastal and Salmon River mountains of Western North America, and the humid areas of extreme Southern Africa and Southeastern Australia are the centers of diversity for the slugs and semislugs. New Zealand, Central America, and the Smoky Mountains of Eastern North America also have native slugs, but there are secondary centers.

Many European slugs have been widely transported by man in the last 200 years. These species are closely associated with human dwellings and agriculture. They give a false impression that slugs are virtually ubiquitous. Slugs are almost everywhere today, but only because of man's activities. Their natural distribution is relatively limited in extent.

WATER AND CALCIUM—THE KEYS TO SLUGDOM

What are the reasons for the naturally limited distribution of slugs? Can we identify any factors in the environment that would tend to "promote" slug evolution? Referring again to the basic biology of land snails, the need for conserving moisture is paramount. If asked to design a paradise on earth for land snails, I would list only a few criteria. First and foremost, there should be plenty of water. This can be in the form of essentially daily showers on open or forested lands. Preferably the frequent rains should be accompanied by a high forest cover that holds moisture and keeps the ground level humidity nearly saturated even if a few days elapse between rain showers. A grassy field will dry out far more quickly than a dense forest, since the canopy of leaves shades the ground litter from the sun's rays. With rain and moderate or warm temperatures, plants will grow quickly and there will be ample snail food available. The mass of decaying leaves and twigs provides a thick carpet on top of the soil that is both a feeding ground and refuge for snails.

A moist forest thus provides water and abundant food for a snail. In addition, the snail needs calcium to build its shell. Many snails cannot use calcium directly, but must assimilate it by eating plants that contain quantities of calcium. Building a shell requires a tremendous expenditure of energy on the part of a snail. Remember that the primary function of the land snail's shell is to retard water loss. Protection against predation is a secondary consideration. If moisture conditions are favorable all the time, then the shell has lost a great deal of its potential value.

The areas where the greatest variety of slugs are found are generally mountains whose slopes lie in the path of prevailing winds coming off large bodies of water. They receive a constant supply of moisture. In addition, most of these areas have volcanic mountains with a minimum of calcareous rock formations available for weathering into soil. Thus the supply of calcium in the soil available first to plants and then to snails is quite limited. It is axiomatic among collectors that snails are more abundant on calcareous soils. Many species are restricted to areas of high calcium availability. The limestone hills of Cuba, Malaya, Borneo, Yucatan, and Puerto Rico harbor thousands of snails, while the intervening calcium-deficient soils and forests are near molluscan deserts.

Individual snails are also affected by calcium levels. The thickness of their shells changes with the availability of calcium. More than 45 years ago an English zoologist, Charles Oldham, showed that juvenile shells of *Arianta arbustorum,* a common European land snail, reached identical size whether fed on a calcium-rich or calcium-deficient diet, but that the weight of the shell averaged 3.54 to 3.85 times as much in the calcium-rich diet specimens. The difference was in shell thickness and calcium content.

Individual snails depend on calcium availability to build a thick or thin shell, but suppose a mutation occurred that would reduce the size of the shell. This would conserve energy use by the snail, since that much less calcium would need to be extracted from its food. On moist mountain slopes water conservation is essentially no problem, so the necessity for the pallial cavity to serve as a retreat for the head and foot and as a storage area for reserve water is absent. Under conditions of adequate moisture supply. any mutations that reduced the thickness or size of the shell would be favored, as would mutations that shortened the pallial cavity. They would be favored in the sense that energy saved in not building a larger or thicker shell and in not growing a larger visceral hump with longer pallial cavity is available for other purposes.

This saving does not come without a price. Loss of the shell and its water conservation function also produces problems in regard to protection from being eaten, in rearranging body organs, and for surviving periods of temporary dryness. These difficulties are partly solved in a number of ways. Many European slugs, particularly some of the species belonging to *Arion,* and some of the Austro-Zelandic athoracophorids are extremely bright in color. While conspicuous in appearance, and thus calling attention to themselves, their slime has chemicals that are highly distasteful to a predator. Many birds and small mammals feed on snails, but very few eat any number of slugs. Many naturalists have reported seeing a young bird grab an arionid slug, only to drop it almost immediately. The bird then acts as if its mouth has been badly burned. Members of the genus *Prophysaon,* found in the Western United States from San Francisco north to Alaska and inland to Idaho, can, like many lizards, amputate their tail when disturbed. Presumably the wriggling tail satisfies the predator's appetite while the rest of the slug crawls away to safety. Indonesian and Philippine semislugs of the Helicarionidae self-amputate their tail when grabbed. The large Cuban snails belonging to the genus *Polydontes* are reported to self-amputate their tail, which will twitch for up to 54 hours after being severed. Even at a snail's pace, this is sufficient time to escape!

Without the burden of a shell, a slug can slip into narrower crevices and hide from a predator far better than a snail whose foot and tail are the same length. Except when stuffed by a recent meal or greatly swollen by eggs, slugs can distort their body shape to an amazing extent, and slither through tiny holes. Indeed, such distortion is a major way in which slugs attempt to retard water loss. A resting slug (Fig. 1*b*) bears little similarity to the same species when it is actively moving (Fig. 1*a*). The four tentacles are withdrawn into the body, the head retracted under the edge of the man-

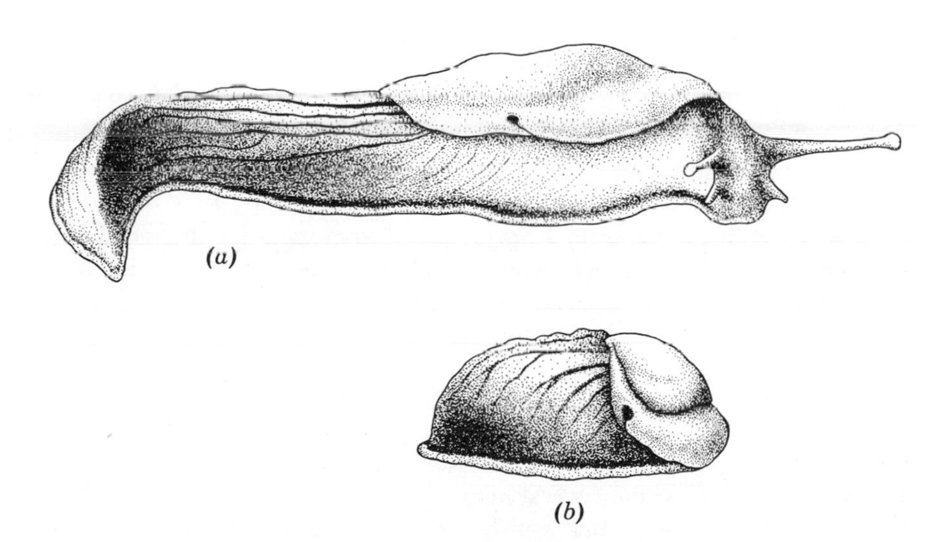

FIGURE 1. A typical European slug shown when crawling (*a*) and resting (*b*). This species, *Milax gagates,* is about 2 to 3 inches in length. After Lovett and Black.

tle lobe, and the tail contracted as close to the back of the mantle as possible. The resting slug may be less than one-fifth or one-sixth its length when crawling. This temporary reduction in length effectively reduces the body surface area exposed to evaporation. Just as a teacup full of water sitting in the sun will evaporate more slowly than the same quantity of water poured onto a dinner plate, so the resting slug will lose water more slowly than one that is crawling.

Crawling itself results in water loss, since the slime trail on which snails and slugs glide consists mainly of water. A thick, very sticky mucus is secreted by the pedal gland at the anterior of the foot, while tiny glands on the sole of the foot secrete a very thin and watery mucus. Both types of mucus function in locomotion. Presumably the balance between the two depends on whether the animal is climbing up a surface, and needs "sticking power," or whether it is traveling in a horizontal plane. B. H. Dainton found that slugs crawling in a saturated atmosphere lost about 17% of their body weight in 40 minutes, while surface evaporation from inactive slugs kept at 45% relative humidity was about 3 to 5% of the body weight per hour. Slugs can absorb water directly through their skin, and Dainton showed by a simple experiment that the water can be absorbed as water vapor. After forcing a 23% body weight loss in a slug, she suspended a slug in a sealed container on a porous platform that was well above the water surface. In about 2 hours the slug had reached its original weight simply by absorbing water from the air.

Experiments early in this century showed that most slugs could stand a 50% body weight water loss and still recover, while some investigators have claimed that recovery in some species can occur after an 80% weight loss. These figures greatly exceed those recorded for most snails. Thus slugs compensate for their increased rate of water loss by movements to wetter crevices, by reducing the surface area exposed to evaporation, by the ability to absorb water vapor through the skin, and by the increased ability to tolerate body weight loss without death occurring.

LAND SNAIL STRUCTURES

One of the major practical problems involved in slug evolution was in regrouping the organs as the visceral hump was reduced in size and eventually lost. Each group of slugs and semislugs has its own peculiarities as to how this change was effected, but a few general patterns exist. Before discussing the ways in which slugs evolved, some of the basic anatomical features of plumonates must be described. The first body region on which to focus is the pallial cavity. For the land snail this provides an empty space into which the head and foot can be retracted. In addition, on its upper posterior surface are situated the heart, kidney, and ureter. Much of the anterior upper surface is heavily crisscrossed by blood vessels, and respiration (Fig. 6*a*) takes place here inside the cavity. The hindgut continues along its upper margin forward to the anus. In advanced pulmonates the ureter parallels the hindgut, finally opening to the exterior alongside the anus. Both of these openings usually are just inside the pneumostome or breathing pore, but occasionally the openings will be slightly separated from each other, although closely clumped together. In addition, most of the upper side of the pallial cavity is highly vascularized for gas exchange. As mentioned previously, the pallial cavity can also serve as a water reservoir. In straight lateral view (Fig. 2*a*) the heart and most of the vascularization cannot be seen, but Fig. 4 on page 222 shows a typical pallial region configuration. Omitted from this drawing are the extensive pulmonary veins that cover much of the pallial surface between the kidney and pneumostome.

The pallial region gives one of the major clues as to why slug evolution is restricted to a few groups of land snails. All land slugs have a closed and rather lengthy ureter that opens directly to the outside of the animal. As in many animals that live on land, one function of the ureter is to resorb water from the excretory products. A major difference between the orders of land snails is whether or not a closed ureter is present and reaches the pneumostome. It is only in the orders Sigmurethra and Systellommatophora, the two major groups with a closed ureter, that slugs and semislugs have evolved. I interpret this to mean that the presence of a water conserving ureter was a preadaptation for the evolution of slugs.

Features of the snail's digestive, reproductive, and muscular systems are designed to enable effective use of the pallial cavity as a space retreat. Figure 2*b* diagrams the snail's digestive system. The mouth, radula, and buccal mass were discussed in Chapter IX. Leading from the buccal mass is a long and slender esophagus that continues along the inner margin of the pallial cavity past its apex and gradually expands into the stomach. The expansion occurs a fraction of a whorl above the pallial cavity apex. The stomach itself continues apically from one-half to two whorls, occupying at least three-quarters of the cross-sectional area of the whorls. Extending from just below and then in narrow strips that run alongside the stomach, then continuing to the very highest point of the shell is the massive digestive gland or "liver" (not shown in Fig. 2*b*). At its apex, the stomach narrows and abruptly reflexes anteriorly as the intestine. This narrow tube passes below and alongside the stomach to just above the pallial cavity apex. There it is complexly coiled before continuing anteriorly as the hindgut to the anal opening at the pneumostome. The long esophagus and hindgut are needed to preserve the empty space function of the pallial cavity. The large stomach is situated well above the pallial cavity apex (PCA), in a region not involved with providing room for head and foot retraction.

Inspection of Fig. 2*c* shows that the reproductive system is organized in part to maintain the empty space area. The ovotestis is embedded in the lobes of the digestive gland well above the stomach. Leading from the ovotestis is a long slender tube, the hermaphroditic duct. This duct carried eggs and/or sperm past the expanded stomach. Between the stomach and apex of the pallial cavity, in the area of intestinal looping, sits the albumen gland, which furnishes the nutritive material for the fertilized egg, plus the spermathecal head, a storage organ for sperm. Leading from these to the genital pore are a series of usually narrow tubes. These are grouped with the hindgut and esophagus to form a compact mass ascending along the inner wall of the pallial cavity. During the period of reproductive activity, the reproductive ducts can become much larger in size, but normally they are slender tubes that do not interfere with retraction of the head and foot. Digestive and reproductive systems are seen to interdigitate tubes and large organs, sharing the upper visceral hump quite efficiently. Both systems include narrow tubes along the pallial cavity. Position and size of the penis are highly variable. In many taxa it is short and reaches barely to the pallial cavity level. In others (as in the diagram) it can be long and slender, reaching to the top part of the pallial cavity. Natural selection for size of the terminal genital organs is not basically concerned with space inside the body, but in distinguishing members of a species from individuals belonging to closely related species.

Yet another system, that of the free retractor muscles, occupies much of the same area. Figure 3 shows the main muscles in side (*a*) and top (*b*) views. A massive set of buccal retractors (BR) pulls the buccal mass and front part of the body backward. Long, slender ocular retractors (OR) and lower tentacular retractors (TR) serve first to pull the feelers into the head region and then assist slightly in retraction of the entire head region. A huge assortment of muscle fibers, collectively called the tail fan, radiates

---

FIGURE 2. Arrangement of basic pulmonate organ systems: (*a*) location and extent of pallial cavity, whose apex is indicated by PCA: (*b*) location of digestive tract—the digestive gland that occupies most of the area above the stomach has been omitted; (*c*) position and extent of genital ducts. The stomach is indicated next to the narrow tube of the hermaphroditic duct. Organs above the pallial cavity apex (PCA) are termed "apical genitalia"; organs below the pallial cavity apex are referred to as "pallial gonoducts." The penis and female organs near the gonopore are sometimes called the "terminal genitalia."

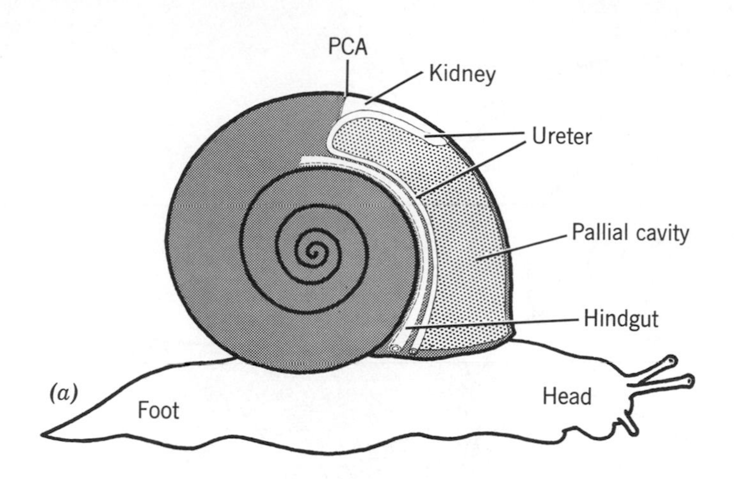
PCA
Kidney
Ureter
Pallial cavity
Hindgut
(a)
Foot
Head

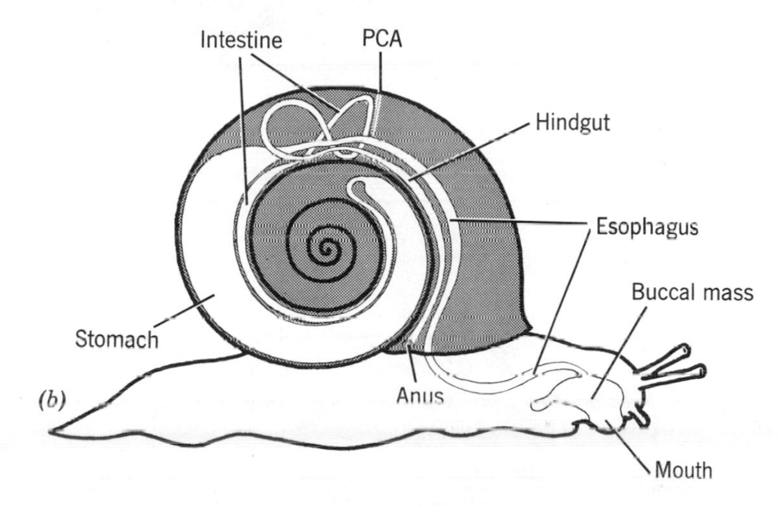
Intestine
PCA
Hindgut
Esophagus
Buccal mass
Stomach
(b)
Anus
Mouth

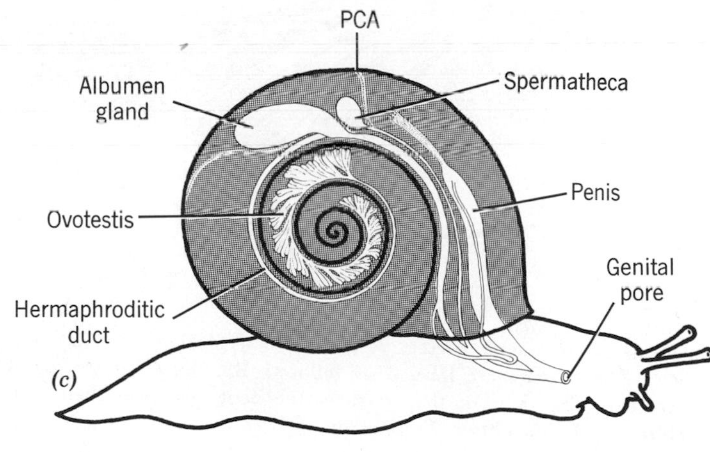
PCA
Albumen gland
Spermatheca
Penis
Ovotestis
Genital pore
Hermaphroditic duct
(c)

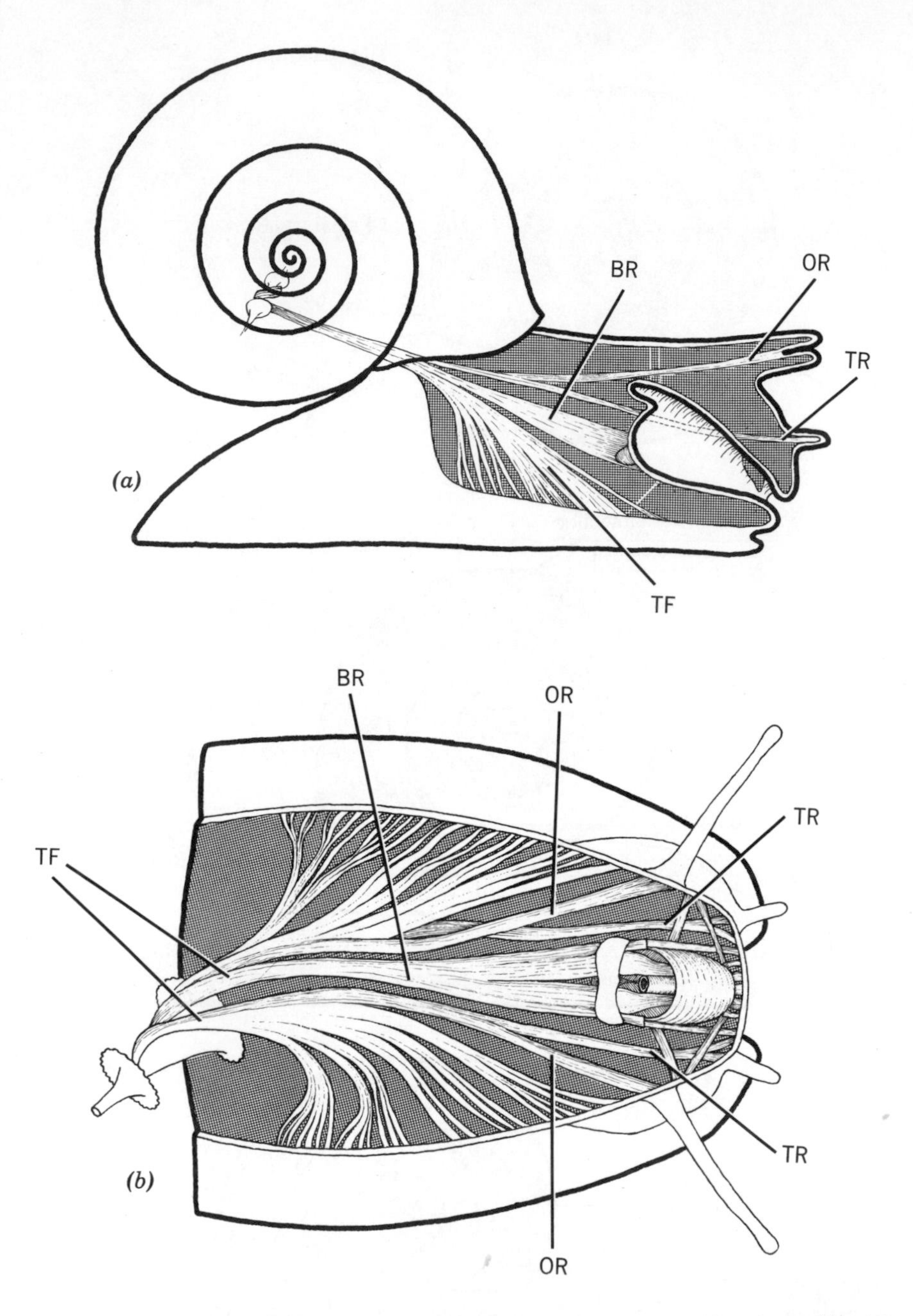

FIGURE 3. Retractor muscle system of a typical snail: (*a*) side view; (*b*) top view. The muscles are labeled as follows: BR, buccal retractor; OR, eye tentacle retractor; TF, tail fan that retracts the foot into the shell; TR, lower tentacle retractor. Modified from Trappmann.

to the sides of the body cavity in the foot and tail region. These are the muscles mainly responsible for actually pulling the head and foot back into the pallial cavity. The free muscles extend usually a whorl or so above the pallial cavity apex, gradually uniting into a single broad muscle called the columellar that is attached to the shell. This is the only point at which the animal is actually attached to the shell itself. The anchor provided by the columellar muscle attachment often is two or three whorls back of the aperture.

CHANGED SLUG ANATOMY

After this survey of the topography found in a snail's body, the problems involved in slug evolution are far easier to understand. Again, by viewing each organ system complex in turn, the road to slugdom is revealed. First we consider the pallial cavity.

In a typical snail the pallial cavity may extend one-quarter to more than a full whorl behind the shell aperture, depending both on the number of whorls and on the length of the snail's foot. When the number of whorls in the visceral hump is decreased, the pallial cavity becomes shorter and broader. Generally the snail's foot may become somewhat shorter and broader so that it can still be withdrawn into the altered cavity.

This pattern of change in the pallial organs is seen in some Pacific Basin Charopidae (Fig. 4). These ground or semiarboreal snails are mostly 3 to 10 millimeters in size. Only a few species have evolved all the way to slugdom. The pallial region of a Samoan charopid which has over five whorls (top) is typical in having a U-shaped kidney with one lobe of the kidney lying next to the hindgut. The two parts of the ureter lie parallel both to each other and to the hindgut. In one lineage (left side of figure) shortening of the pallial cavity is compensated for by rotation of the kidney away from the hindgut. There is minimum shortening of the kidney, and a tendency for the ureter arms to spread apart into a wide V-shaped arrangement. This is the situation in two New Guinea genera, *Pilsbrycharopa* (left center) which has slightly less than four whorls, and *Paryphantopsis* (bottom left), which has only three whorls.

Other species have solved the same problem in a different way. A Caroline Island genus with $2\frac{3}{4}$ whorls (right center) has the kidney shortened and broadened, but the ureter arms are squeezed more closely together rather than diverging. The New Zealand *Flammulina* (lower right) which has $3\frac{1}{4}$ whorls, has the kidney shortened and broadened, the ureter arms diverging, and a small "bladder" developed just beyond the opening of the ureter. Whether this "bladder" functions more in water resorption than as a sac

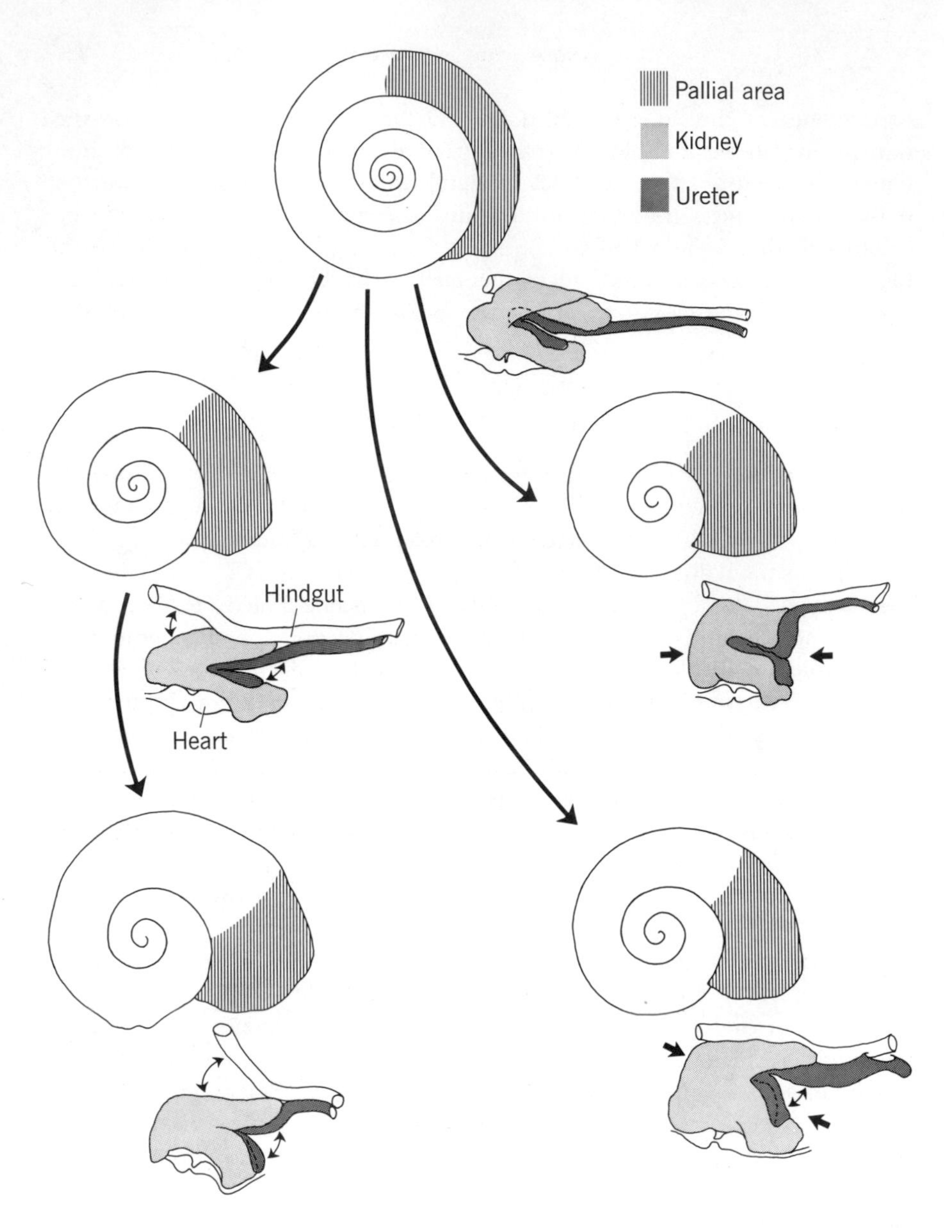

FIGURE 4. Pallial area changes in Pacific Basin Charopidae. Diagrams are based on a Samoan genus (top); the New Guinea genera *Pilsbrycharopa* (left center) and *Paryphantopsis* (left bottom); a Caroline Island genus (right center); and the New Zealand *Flammulina* (right bottom). Modified from research by the author.

for holding waste products is unknown. The problems of organ arrangement and shape when space reduction occurs can be solved in quite different ways within the same family.

Even more dramatic changes occur in members of the Helicarionidae. This dominant family of the Southeast Asian tropics has many sluglike taxa. The patterns of change in several Thailand genera are shown in Fig. 5. *Durgella* has a shell with four whorls and is a reasonably typical snail. The pallial region (top) has an elongately triangular kidney whose posterior end touches a loop of the intestine. There is a reflexed ureter, one part of which lies next to the hindgut, and a typical heart and venous systems. In *Megaustenia,* which has only $2\frac{1}{3}$ whorls (center right), the shortened pallial cavity has been compensated for by the kidney extending partly above and below the intestinal loop. *Cryptaustenia* (left center), which has almost four whorls, has the kidney reflexed into a U-shaped organ, with the two lobes unequal in length. This is an early stage in the evolutionary sequence that culminates in the slugs *Austenia* (lower left) and *Muangnua* (lower right). Both of these have the lobes of the U-shaped kidney equal in length. In *Muangnua,* where the shell is flatter and the pallial cavity further reduced in size, the ureter lies partly on top of the kidney, rather than alongside of it as in the other genera.

Specimens of *Durgella* and *Megaustenia* can withdraw the head and foot completely into the shell. In *Cryptaustenia* the pallial cavity is too small to hold the entire foot, which is greatly elongated posteriorly and may be capable of self-amputation as a defense mechanism. In *Cryptaustenia* the pallial cavity has started to shorten posteriorly with the initial folding of the kidney a result of this change.

Once the pallial cavity is too small for the foot and head, further shortening and compaction can proceed quite rapidly. This does create a potentially major problem, since the upper surface of the pallial cavity is used in land snails for respiration (Fig. 6*a*). As this area decreases in size the surface available for gas exchange would become inadequate. The snail might smother to death unless other regions provided sufficient surface area for gas exchange. Generally this happens by venous vessels invading the mantle edge and the latter organ developing a lobe of tissue that extends forward along the neck of the snail, and/or various flaps that extend backward over the shell surface. Most of the species in Fig. 8 show extensive development of such accessory breathing areas. In time, these mantle extensions can completely cover the shell surface. *Austenia* (Fig. 6*b*) has a fair portion of the shell exposed. In *Ranfurlya* (Fig. 6*c*) a caplike remnant of the shell is visible, but in *Muangnua* (Fig. 6*d*) and other true slugs the shell is completely covered. In intermediate situations, these shell flaps can be extended or con-

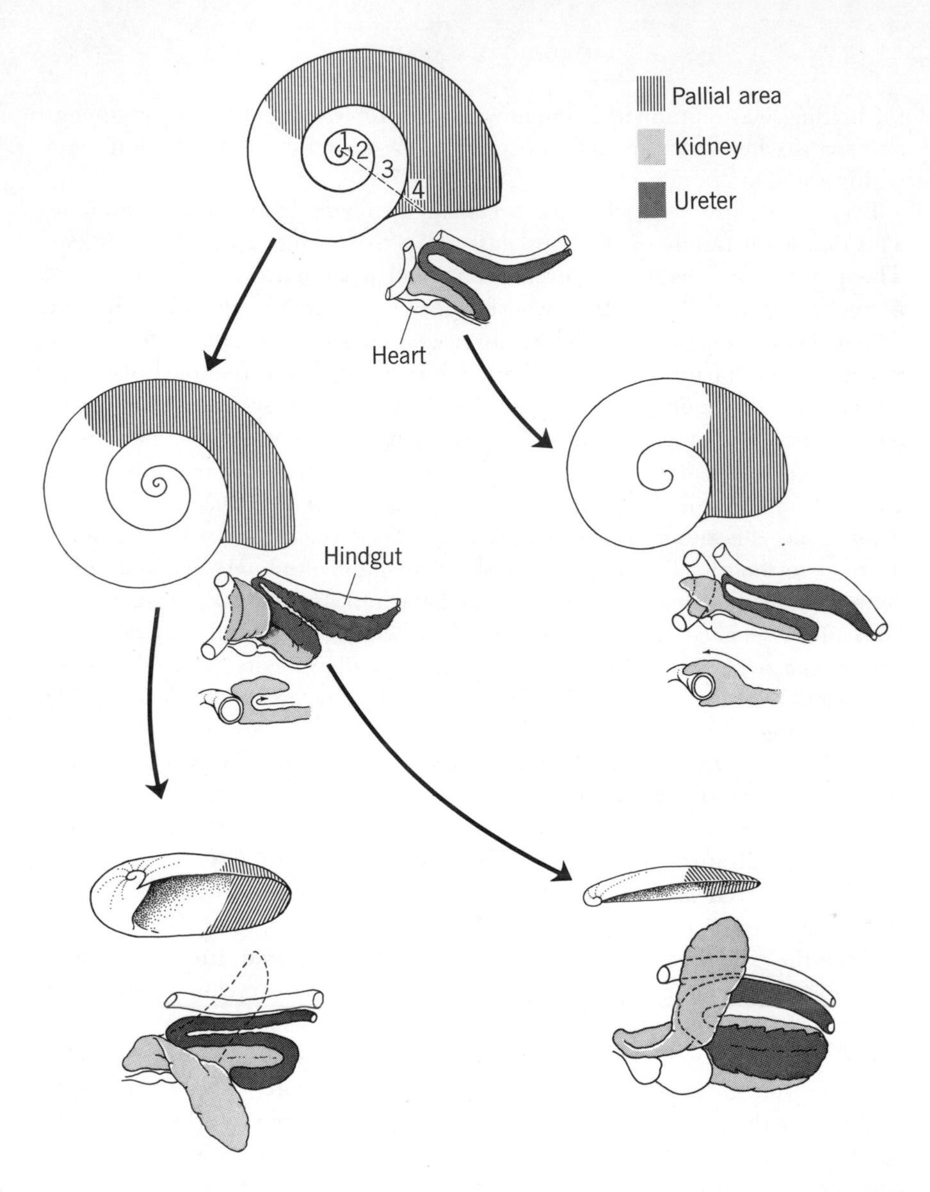

FIGURE 5. Pallial area changes in Thailand Helicarionidae. Diagrams are based on the genera *Durgella* (top); *Cryptaustenia* (left center); *Megaustenia* (right center); *Austenia* (left bottom); and *Muangnua* (right bottom). Based on a 1966 publication by the author.

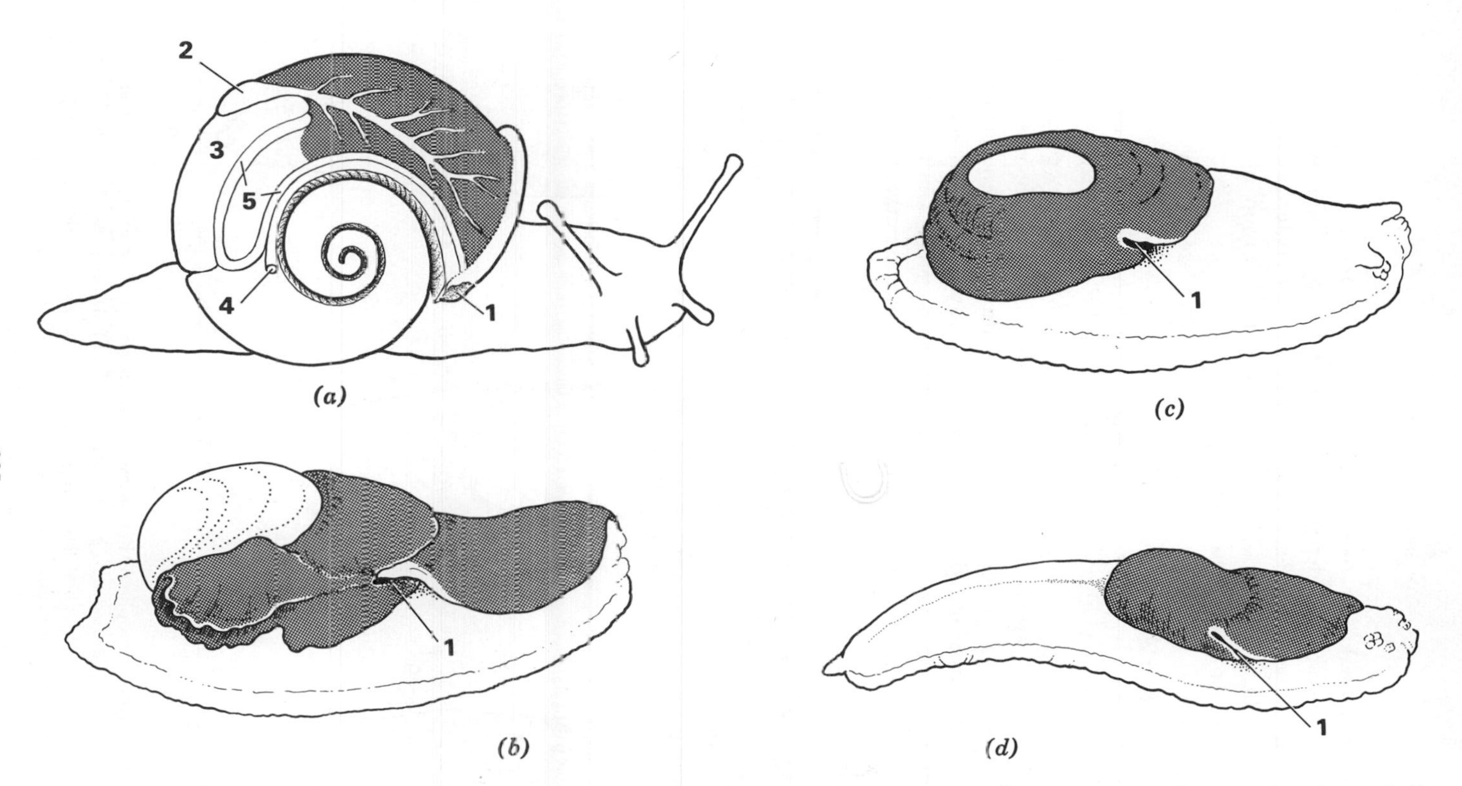

FIGURE 6. Respiratory surfaces in snails and slugs: (*a*) *Helix,* with shell removed; (*b*) *Austenia;* (*c*) *Ranfurlya,* with tip of shell still exposed; and (*d*) *Muangnua,* with shell completely enclosed by the shell laps. Shaded area represents actual surfaces involved in respiration. Identified structures are the pneumostome or breathing pore (1), heart (2), kidney (3), hindgut (4), and ureter (5).

tracted by the animal. The two Queensland *Helicarion* shown in Plate 10, center right and lower right, demonstrate the extremes quite effectively.

These are only two examples of the ways in which pallial structures have been altered to fit reduced pallial cavity space. Each group of slugs and semislugs has its own equivalent story.

Becoming sluggish involves two other changes that are equally dramatic and important. In typical land snails (Fig. 2*b*) the stomach sits inside the visceral hump well above the pallial cavity. The esophagus conveys food from the buccal mass to the stomach, preserving the empty space function of the pallial cavity. When that space function is no longer necessary, the esophagus is also dispensable. Hence one of the characteristic changes in semislugs is for the esophagus to become dramatically shortened and the stomach to shift anteriorly down into the body cavity. In early stages of slug evolution, such as the New Zealand Charopid genus *Ranfurlya* (Fig. 7*a*), the stomach is partly in the visceral hump and partly in the body cavity.

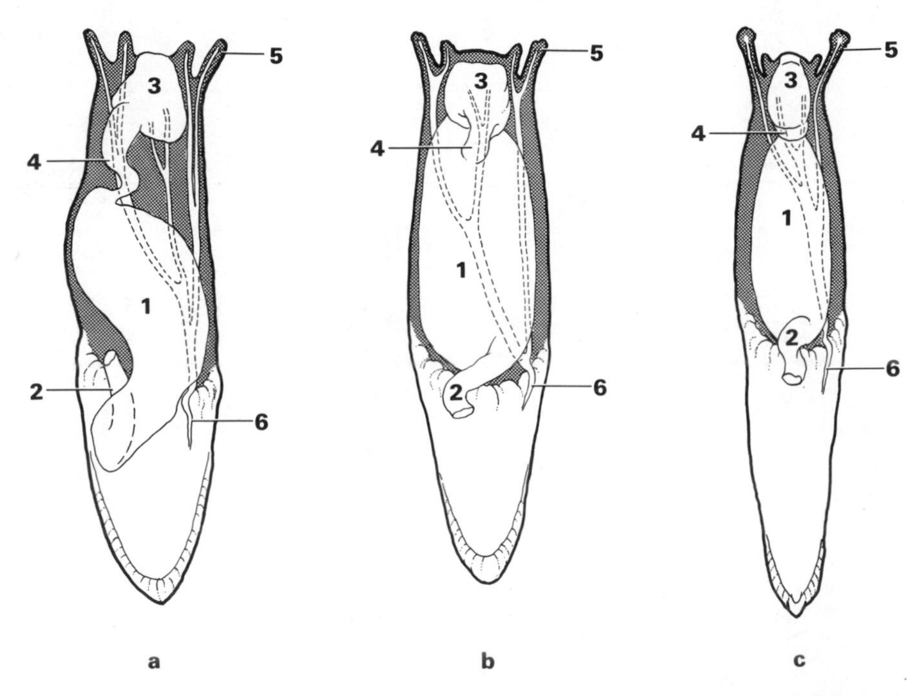

FIGURE 7. Pattern of esophageal reduction, stomach shift, and free muscle loss in sluglike taxa: (a) *Ranfurlya;* (*b*) *Austenia;* and (*c*) *Muangnua.* Identified structures are the stomach (1), first loop of intestine (2), buccal mass (3), esophagus (4), eyestalks (5), and columellar muscle (6).

In later stages such as *Austenia* (Fig. 7*b*) and *Muangnua* (Fig. 7*c*), the stomach lies completely within the head and foot region. The visceral hump is reduced to a small caplike structure that holds only the apical genitalia, digestive gland, and intestinal loops. This is the common pattern in European slugs such as *Geomalacus* (Fig. 8*d*). Eventually the visceral hump completely disappears and a slug such as *Veronicella* (Fig. 8*f*) or *Pseudaneitea* (Fig. 8*e*) results. At the same time the free muscle system becomes dramatically reduced. No longer is an elaborate tail fan (Fig. 3) needed to retract the head and foot. Generally only a small buccal retractor, tentacular retractors, and a penial retractor muscle remain of the free muscle system. These still attach to the remnant of the columellar muscle (Fig. 7).

Evolution of a slug results from natural selection under conditions of plentiful moisture and scarcity of calcium. The pallial cavity is no longer essential as a space into which the head and foot can be retracted. Subsequent reduction of both the visceral hump and shell occurs. Shrinkage of space in the pallial region results in altering the shape and relative position of various organs. Eventually the stomach will shift down into the body cavity with the esophagus becoming extremely short and the free muscle system reduced to a remnant.

SLUG DIVERSITY

These changes have happened in a number of lineages in many parts of the world. Figure 8 illustrates a few of the many slug taxa. They are particularly common on the south slopes of the Himalayas, continuing into the tropical forests of Southeast Asia and Indonesia. *Cryptaustenia* has most of the shell exposed, but the enlarged foot means that the mantle lobes are necessary to provide adequate respiration. In *Austenia* (Fig. 8*a, b*) the visceral hump and shell are much reduced, with a quite large head-shield mantle extension. *Girasia* (Fig. 8*c*) look virtually identical externally, but have quite distinctive anatomy. They are not closely related.

A few slugs show no trace of a visceral hump at all. The South Pacific athoracophorid slugs (Fig. 8*e*) are fungus feeders whose slime network resembles the venation of a leaf. Currently there is much controversy and little detailed knowledge concerning their ancestors and relationships. Members of the family Veronicellidae (Fig. 8*f*) have a smooth surface, no trace of any pallial cavity, and are unrelated to the main groups of land snails. They, together with a carnivorous group, represent an independent colonization of the land from marine ancestors. Originally veronicellids were common in Africa, South America, and Southeast Asia, but today species have been distributed by man through most tropical areas of the world.

LAND SLUGS

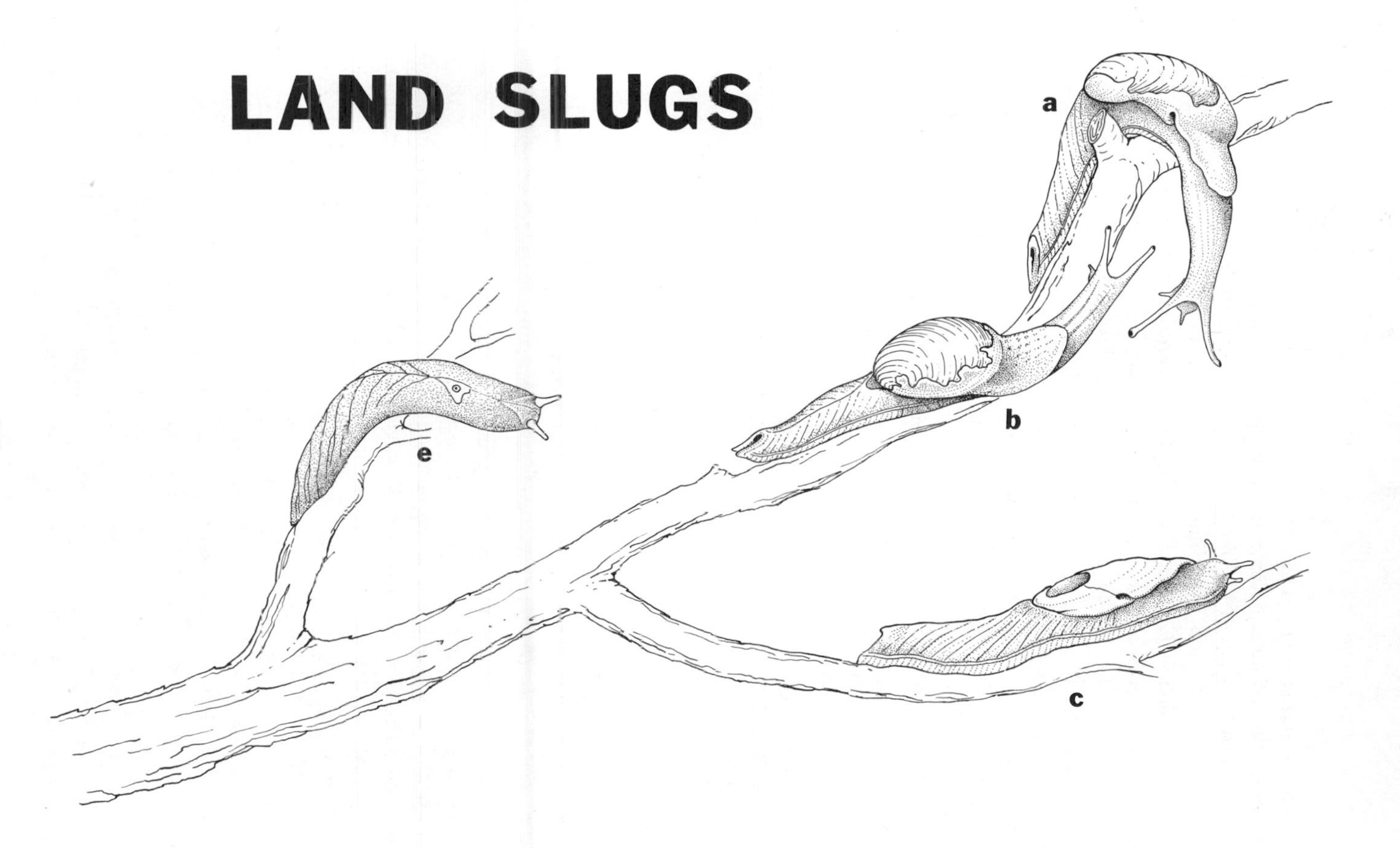

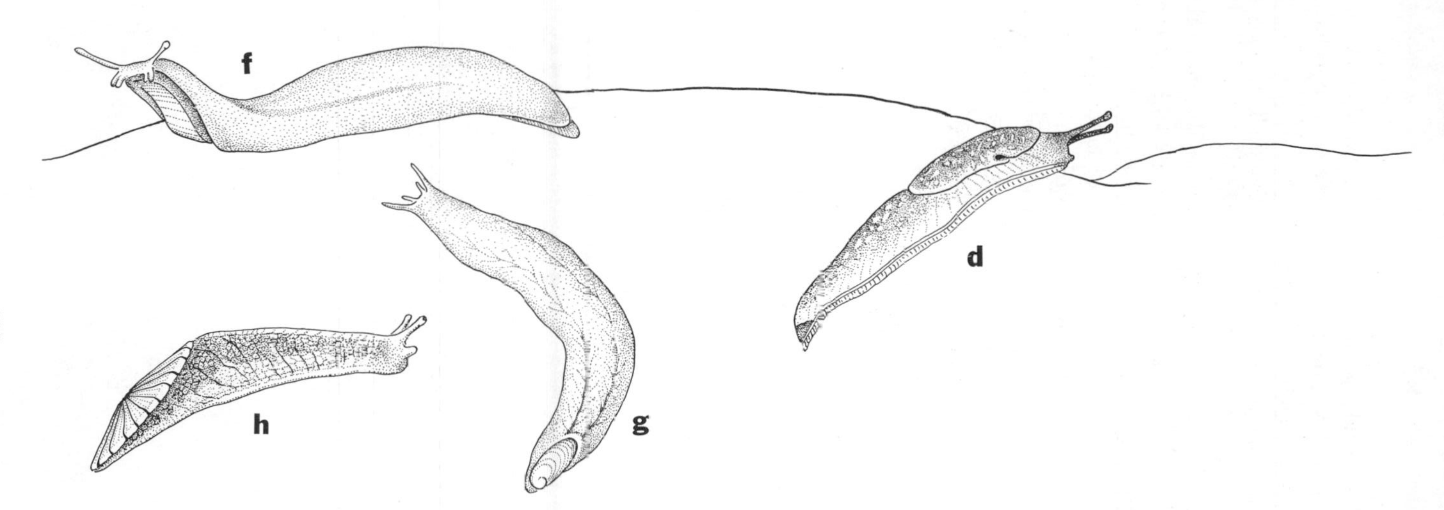

FIGURE 8. Representative slugs and sluglike species. Not drawn to scale. The genera are (*a, b*) *Austenia* from northwest India; (*c*) *Girasia* from the same region; (*d*) *Geomalacus* from Ireland; (*e*) *Pseudaneitea* from New Zealand; (*f*) *Veronicella* from Mexico; (*g*) *Testacella* from England; and (*h*) *Apera* from the Union of South Africa. *a–c* after Godwin-Austen; *d* and *g* after Taylor; *e* adapted from Burton and slides of living examples; *f* after Fischer and Crosse; *h* after Collinge.

In two areas of the world carnivorous slugs have evolved from shelled ancestors. The lengthened body of carnivorous snails is reflected in the appearance of these slugs. Both the European *Testacella* (Fig. 8*g*) and South African *Apera* (Fig. 8*h*) have the remnant of the shell and visceral hump perched on the back of the tail.

This chapter has presented only a small sample of the snails that became sluggish and has given a brief look at the problems and patterns involved in their evolution. Although most people look upon slugs as loathsome garden plant ruining pests, they represent the acme of land snail evolution and are fascinating, frequently colorful, but little known creatures. The recent book *Terrestrial Slugs* by Runham and Hunter is an excellent introduction to their biology, structure, and functioning. It essentially starts where this chapter leaves off. I highly recommend it for further reading.

XIII
Why Who Lives Where

Walruses are found in the high Arctic, but not near the South Pole. Penguins live in the southern hemisphere, while auks dwell in the Arctic regions. Elephants are native to Africa and Southeast Asia, kangaroos to Australia, and pandas are restricted to Southwestern China. These "facts of distribution" are much better known than the equivalent ones concerning mollusks. Even few malacologists recall immediately that clausiliid land snails are common over Eurasia and in the Andean regions of South America, only barely reach New Guinea, have single species on Haiti and Puerto Rico, and are absent from the rest of the world. Pleurotomarian slit shells are found on rocky cliffs in moderately deep to deep ocean waters, whereas most cowries live in the tidal zone or in relatively shallow reef waters.

Every group of organisms shows its own peculiarity of distribution. These are determined partly by physiological factors which set the limits of tolerance. These limits can be the extremes of such items as temperature, salinity, and presence of trace minerals in the ocean, or moisture and temperature on land, between which a species can live and successfully reproduce. Equally important are biotic factors, such as the presence or absence of a needed food item, the absence of a fish host to complete a unionid life cycle, the presence or absence of a highly effective predator or parasite, or the presence or absence of closely related species with which individuals must compete for food, space, or some other item that is essential to life. Often the current distribution of an organism is limited by the presence of a physical barrier that prevents further spread. This normally will be dry land to a marine or freshwater organism. Land dwellers are limited either by a body of water or a zone of unfavorable climate and vegetation.

All present distributions are the result of cumulative historical processes, just as living species are the result of cumulative evolutionary events from past eras. Because organisms evolved in different places at different times, show different limits of tolerance, have different capacities for dispersal to new areas, and react differently with other organisms, their current distributional patterns are far from being identical. We know that many species can live quite successfully in parts of the world far removed from their native habitat. Basic food plants—potatoes from South America, wheat from the Near East, corn from Central America, and bananas from South Asia—are examples of how man has purposely introduced organisms from one place to another. Other introductions were accidental or less successful. Virtually every slug that chews on a garden plant in Eastern North America was accidentally imported from Europe. English sparrows, starlings, house mice, Norway rats, ginko trees, and many garden plants were introduced to the United States from elsewhere. Prickly pear cactus to Australia and Africa, the Colorado potato beetle to Europe, and the muskrat to Europe are reverse migrants.

The reasons why these introductions have succeeded are varied. Often there was no natural enemy present in the new area, no organism lived in the same way, or the immigrant was aided directly by man. Thus the introduced species could be spared adverse pressure by a predator or parasite, had no equivalent organism to compete with for food, space, and other niche factors, or was sheltered and protected by man, respectively. Less obvious is the fact that many areas of the world have equivalent climates and types of vegetation cover. Forest and field slugs of Europe found the forests and gardens of Eastern North America the same in terms of moisture and copious vegetation cover. Thus 13 of 16 slug species now living in Northeastern North America are from Europe. The prickly pear cactus found the arid areas of Australia enough like the arid New World areas that it flourished and became a serious pest until insects and cactus diseases were imported to bring it under control. Introduction of species from areas with similar climates and vegetation is relatively easy to undertake, but if the introduced organisms subsequently become pests, then control may be very difficult.

Because widely separated areas can have essentially identical climates, the same type of vegetation, and thus virtually identical opportunities for "making a living," species that closely resemble those in another area have evolved. Every tropical forest has insects that mimic dead leaves, twigs, or even bird droppings. Usually the "look alikes" from the widely separated areas are not related to each other, but have reached the same adaptations independently through "convergent evolution." One of many land snail examples is shown in Fig. 1. *Leptarionta trigoniostoma* (left) is a helmintho-

FIGURE 1. Convergent evolution in tree snails. Left is the 1-inch-diameter *Leptarionta trigoniostoma* from Guatemala. Right is *Papuina boivinii* from Ysabel Island, Solomon Islands.

glyptid land snail from Guatemala, while *Papuina boivinii* (right) is an unrelated camaenid from the Solomon Islands. Both species are tree dwellers, virtually identical in size, shape, and color pattern, but evolved separately from quite dissimilar ancestors.

Biogeography is the science that attempts to delineate and understand the distributions of animals and plants. Often it has been divided artificially into descriptive, historical, and ecological aspects. In recent years the emphasis has been more and more on theoretical and experimental biogeography, with a fusion of ecology and biogeography for the study of diversity. It is not enough to map out distributions of many species, find common or "average" patterns within a group, and divide the world into a series of biotic "realms, regions, and provinces."

Other questions are being investigated. What have been the major historical events that determined present distributions? Man's introductions show that numerous species can live quite well in regions of the world where they did not occur before man carried them about. Why didn't they get there themselves? These concern large scale, long term events.

Some questions focus on local or quick happenings. How far and how quickly will individuals of a species disperse? At what stage of their life history? What happens when only a few of the many different animal and plant species found on a continental area reach a small, isolated island? How often does a new colonizing land dwelling species reach such a group as the Hawaiian Islands? How can such "overseas" dispersal take place? Very recently biogeographers have started to investigate the nature of local diversity.

These are a few of the questions with which biogeographers concern themselves. Data are drawn from many disciplines and from many different groups of organisms, then integrated to try to map out not only the broad patterns of modern biogeography and to interpret changing patterns through time, but also to understand the reasons behind biogeographic changes.

One of the major lessons to be learned from the study of biogeography is the extent to which change is a normal pattern of events. Populations of insects or lizards on tiny islands can be established and then become extinct in the space of a few weeks or months. Species of birds or mammals can be wiped out by man in a few decades. In the same length of time other species can populate a continent. The European starling was successfully established in Central Park, New York City in 1891. By 1953 it was found in vast numbers throughout the United States. On a slightly longer time scale, major changes in climates and vegetation cover occur. Over the last 100,000 years glaciers came and went several times over much of Europe and North America. There are even long term trends. At the start of the Tertiary, London and New York were surrounded by tropical forests, and even in February ocean temperatures off Seattle averaged more than 75°F. Since then the climate has cooled off considerably, with the severe winters and frigid waters found in these areas today harboring quite a different variety of life.

Here we can examine briefly only a few aspects of biogeography: basic divisions, the grand patterns of change, island colonization, and diversity. Because this is a complex subject, examples have been taken from the biogeography of terrestrial organisms. Terrestrial and freshwater biogeography are far better understood than marine biogeography. The nature and variety of barriers on land are more obvious. In addition, first level biogeographic studies depend on accurate systematic studies and detailed accounts of distributional limits. Because mammals, birds, and flowering plants have been studied more intensively than other groups of organisms, the early biogeographers drew their conclusions from an analysis of either birds and mammals or plants.

BASIC ZOOGEOGRAPHIC DIVISONS

Work in the 1850s by Ludwig Schmarda on mammals and P. L. Sclater on birds form the basis of a great synthesis Alfred Russell Wallace published in 1876 entitled *The Geographical Distribution of Animals*. These scientists recognized that very similar types of birds and animals extended over great areas, and that in a few places there were sharp changes in the kinds of animals over a very short distance. Their division of the world into six main zoogeographic regions still forms the basis for much descriptive biogeography:

PALEARCTIC REGION.	Europe and Asia north of the Himalayas, Africa north of the Sahara
NEARCTIC REGION.	North America, temperate and desert parts of Mexico
ETHIOPIAN REGION.	Africa south of the Sahara, Madagascar, and adjacent islands
ORIENTAL REGION.	India, Ceylon, Bangladesh, Southeast Asia, South China, Malay Archipelago east to Celebes
AUSTRALIAN REGION.	Australia, New Guinea, New Zealand, and often the Pacific Islands
NEOTROPICAL REGION.	Tropical and subtropical Mexico, Central America, South America, West Indies

Some prefer to unite the very similar Palearctic and Nearctic regions into a single Holarctic Region. Literally hundreds of papers have been written discussing the exact limits of such regions and how to treat transitional areas (places where animals characteristic of two regions mingle together in varying proportions) such as Central America and parts of Indonesia. In time the recognition that distribution of *organisms* is more important than delineation of geographic *areas* quieted much of this discussion and helped channel biogeography into more fruitful investigations.

For a long time biogeographers focused on two very puzzling problems. First, the major distribution patterns shown by organisms such as mammals, snails, plants, and amphibians are not the same. New Guinea, for example, has Australian frogs and mammals, but its plants, snails, and insects are mostly Indo-Malayan groups belonging to the Oriental fauna. Secondly, many organisms show *disjunctive* distribution patterns. Disjunctive distributions are where organisms live in two or more separated regions, but are absent from intermediate areas. Such distributions are quite common among plants, snails, and insects, moderately frequent in reptiles and amphibians, and very rare among mammals. The only conspicuous disjunctive mammalian distribution is that of the tapirs, who live in both tropical America and Southeast Asia. Since fossil tapirs are known from many places in the northern hemisphere as recently as the Pleistocene, their current disjunctive distribution resulted from tapirs becoming extinct in the northern intermediate regions. Even among the bats, there are few very conspicuous disjunctive distributions. The sheath-tailed bats (Family Emballonuridae) are found in all tropical regions, but absent from temperate regions, while the hollow-faced bats (Family Nycteridae) range from South Africa to Arabia, then from Burma through part of Indonesia.

In contrast, the distribution of land snail families shows a significant percentage of disjunctive distributions. Among the families mentioned on earlier pages, the following show major geographic gaps between areas of abundance:

HELICINIDAE.	Neotropical, Oriental, part of Australian
POTERIIDAE.	Neotropical, some Pacific Islands
POMATIASIDAE.	Neotropical, Ethiopian, few in Palearctic
CLAUSILIIDAE.	Palearctic, Oriental, Neotropical
STREPTAXIDAE.	Ethiopian, Oriental, Neotropical
RHYTIDIDAE.	South Africa, Australian, some Pacific Islands
BULIMULIDAE.	Neotropical, some Pacific Islands
CHAROPIDAE.	Australian, South Africa, Neotropical
CAMAENIDAE.	Oriental, Australian, Neotropical

A very few helicinids, pomatiasids, and bulimulids, for example, reach parts of North America, and there are other minor "spillovers" from regional boundaries, but the major disjunctions in the distributions outlined above are real and significant. There are other families of land snails and many freshwater snail and clam families, in addition to numerous marine mollusk groups that show disjunctive distributions.

Insects and plants show even greater degrees of family level distributional gaps. This has been particularly well studied in relation to the flowering plants. Robert Thorne has calculated that 78% of the 324 flowering plant families and roughly one-quarter of the living plant genera show "intercontinental discontinuities" in distribution.

There is thus a great difference in distributional patterns, from mammals that show virtually no family level discontinuities, to amphibians and snails with moderate numbers, to insects and plants with clear majorities of the families occurring in two or more widely separated areas.

Attempts to explain these differences still lead to heated arguments. In the early part of this century, a school of zoogeographers tried to explain such discontinuities by "building hypothetical land bridges" across ocean deeps to populate isolated islands with a few insects or snails. This approach has been thoroughly discredited. Other biogeographers, primarily students of vertebrates living in the northern hemisphere, argued that the continents had been fixed in their current positions through geologic time. Organisms now found only in South Africa and Australia, for example, formerly must have been present in Eurasia and then been replaced by other groups, much as the tapirs used to live in North America and Europe, but today are found only in Neotropica and Malaysia. This view of basic distribution was strongly championed in W. D. Matthew's *Climate and Evolution* published in 1915. The distribution of recent and fossil mammals was convincingly explained by Matthew on the basis of continental permanence, and dispersals of higher vertebrates from the northern to southern continents over temporary land connections, such as the Bering bridge from Siberia to Alaska, through Panama from North to South America, and through Indonesia from Southeast Asia to Australia. A more recent biogeographer, P. J. Darlington, has summarized the basic dispersal of vertebrates in two diagrams reproduced here as Figs. 2 and 3.

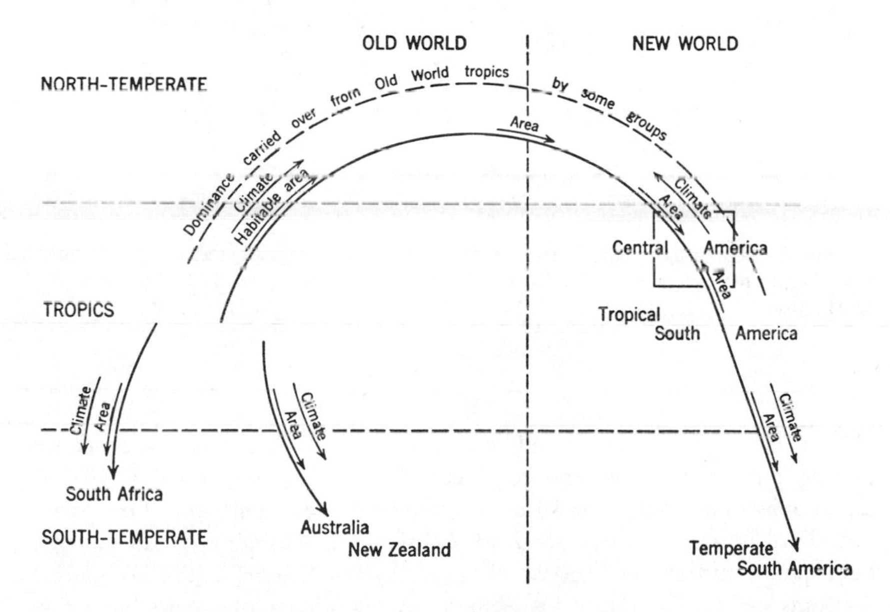

FIGURE 2. Main dispersal patterns of the vertebrates. It is hypothesized that most groups originated in the Old World tropics, with subsequent dispersal to other areas. From Darlington, 1948; reproduced with the permission of *The Quarterly Review of Biology* and the author.

FIGURE 3. The three main dispersal routes of the vertebrates, shown on a double orthographic projection of the world. Courtesy John Wiley & Sons, Inc. and P. J. Darlington.

Coincidentally, 1915 saw publication of a work by Alfred Wegener suggesting that in the past the continents were originally united in a single mass. Subsequently they broke up and slowly drifted apart into their current positions. This theory was received skeptically by residents of the northern hemisphere and students of vertebrates, but much more enthusiastically by residents of the southern hemisphere and students of invertebrates and plants. During the 1960s dramatic geologic evidence accumulated to show not only that "continental drift" had occurred, but to propose detailed timing and sequential occurrence of the splitting. Current information suggests that South America and Africa separated in the Jurassic or early Cretaceous, starting the breakup of a great southern continent called "Gondwanaland," which consisted of South America, Africa, Antarctica, India, Australia, New Guinea, New Zealand, and various satellite islands such as Madagascar and New Caledonia. The exact sequence of this breakup is still being debated, but the times at which pieces split off vary from 50 to 100 million years ago.

Proof that the southern continents and larger islands were connected up to about 100 or 150 million years ago and then gradually split and drifted into present positions is forcing a major revaluation of biogeographic ideas and systematic relationships. The degree of impact that continental drift theory has on systematics and biogeography depends on the time of origin for the organisms being studied. Most families of living mammals, for example, evolved in the Eocene to Oligocene (26 to 50 million years ago) and their distributions date from well after the breakup of the continents. On the other hand, primitive birds such as the ratites (kiwi, emus, ostriches, cassowaries, moas, tinamous) and penguins evolved prior to the Eocene and may have been partly dispersed by continental drift. Flowering plants originated way back in the Mesozoic. Probable flowering plant leaves have been reported from Triassic strata and wood fragments from Jurassic rocks, which is prior to the common appearance of flowering plant fossils in Cretaceous rocks. With the great number of family level disjunct distributions, 78% of living families, students of plant distribution can be expected to find many distributions that might have resulted from continental drifting. However, we must now guard against the natural tendency to automatically explain all major discontinuities in current distributions by continental drift. Robert Thorne reports that 186 of the 276 flowering plant families recorded from South America and Africa occur on both continents. Yet his analysis of their relationships resulted in concluding that either common derivation from Antarctica or rare overseas dispersal would account for the striking floristic similarities between Africa and South America.

Having this background information on the problem of distributional discontinuities and the question of directional movements of organisms, what can be said concerning land snails and continental drift? Several families (see p. 236) show major distributional discontinuities, but unfortunately we do not have enough information concerning exact relationships between the species and genera found in different areas to say what their origins and patterns of movements through time might have been. It will take detailed studies by many workers before we know enough about their relationships to make biogeographic judgments. Clausiliids from South America and Southeast Asia do not differ in major features according to recent work by Loosjes, and my own studies suggest that Australian and South American camaenids show the same patterns of anatomical variations. But this is only fragmentary data.

ISLAND COLONIZATION

The comments made in the preceding section concern land areas that have been interconnected at one time or another. Although they may have separated millions of years ago, the distinction between such *continental* land areas and *oceanic islands,* which have never been connected to another bit of land, is of fundamental importance to biogeographers. The former regions could have received at least part of their animals and plants by dry land passage. A reasonably balanced sample of prey and predators, flowers and pollinators, producers and reducers, arrived together. The ways of living were in balance, and thus the opportunities for wild evolutionary experimentation were lowered through the "checks and balances" of ecological interactions.

Oceanic islands present a different picture. After the islands were first elevated above the sea, pure chance determined which organisms arrived when. At various times, partly drowned organisms on a floating log, fern spores carried in air currents, weak flying insects swept from a land mass by breezes, seeds in a bird's stomach, or even snails clinging to leaves wrenched loose from trees by a cyclone could land on the new island. Once plants were established, insects, snails, spiders, and birds could survive, multiply, and eventually experiment with different ways of living.

A snail that fed on decaying leaves, for example, might on a continental area never successfully "climb a tree" to feed on algae and lichens. Other snails might live in that manner, or predators might exist that would pluck every venturesome snail from the tree trunk. Both competing snails and predators could be absent from the "oceanic island." The algae feeding niche could be vacant. Gradual experimentation by the descendants of snails lucky enough to arrive early could produce highly unusual varieties. The same principles apply to plants, insects, and birds.

In fact, the pattern of organisms living on oceanic islands is to display exactly that type of variety. The finches on the Galapagos, the honey-creepers of Hawaii, and the moas of New Zealand are examples from the bird world. Bulimulid snails on the Galapagos, and achatinellid snails and fruit flies on the Hawaiian Islands are other groups that have undergone major "adaptive radiations" on isolated islands. For many years biogeographers were quite reluctant to accept the idea that islands more than 2000 miles from a major land mass could be populated by "overseas" dispersal. It seemed so improbable that any successful colonization could occur over such distances.

The Hawaiian chain, for example, lies a minimum of 2400 miles from the American mainland. The oldest of the Hawaiian group that is still above water, Midway Atoll, is less than 25,000,000 years old. The youngest of the Hawaiian Islands, the "big island"—Hawaii—is perhaps only 1,000,000 years old. Despite their young age and isolated geographic position, the fauna and flora of Hawaii are world famous for their variety. Conservative estimates of the Hawaiian species given by Elwood Zimmerman amount to about 3800 insects, 1000 land snails, and about 1729 flowering plants. This does not mean that 6500 species reached the islands independently. Probably no more than 275 "pregnant female" insects, 25 different land snails, and 256 plants could have served as the basis for the present diversity. A single successful immigrant might, in time, have given rise to 100 to 300 species. If the oldest Hawaiian island first became land 25,000,000 years ago, then *one successful colonization by an insect, snail, or plant every 45,000 years* would have been sufficient to provide the ancestors needed for today's diversity. Even if the oldest land area in the Hawaiian chain turned out to be only 5,000,000 years old, then one colonization every 9000 years would be sufficient.

How do they arrive across water gaps? By rare accidents. S. Carlquist analyzed the Hawaiian plants. Probably three-quarters could have been carried by birds, mostly as seeds in the bird's stomach or stuck to feathers. A few might have traveled in mud dried onto a bird's foot. More than one-fifth probably drifted in on ocean currents. The Hawaiian snails were probably carried by birds or blown in by cyclonic winds.

Similar stories can be told for many other oceanic islands. The land snails of St. Helena and the Canary Islands in the Atlantic, of the Balearic Islands in the Mediterranean, Mauritius and Reunion off South Africa, the Galapagos and Juan Fernandez off Western South America, and virtually all the high islands of the Pacific Ocean, from the Marianas and Palaus to Marquesas and Easter Island, present a dazzling variety of evolutionary experiments. Plants, birds, and insects are often equally varied.

Because of their spectacular nature, island organisms have been favorite subjects for study by biologists. Many of Darwin's ideas concerning evolution were sparked by his visit to the Galapagos. Unfortunately, the opportunities for modern studies are rapidly diminishing. While the unbalanced nature of the biota gave great opportunities for evolutionary experiments, the lack of competition and generally favorable conditions resulted in delicately balanced communities that can be easily disturbed. On the Pacific Islands, for example, introduced ants have destroyed literally thousands of insect and snail species. Introduced plants have replaced much of the native vegetation, which in turn led to extinction or approaching extinction of numerous species and genera.

At times man's own introductions have created problems that led to rash actions. In Hawaii, a 4 to 6 inch long land snail introduced from Africa, *Achatina fulica* (Fig. 4*a*), is a pest in gardens and scrub areas, feeding on ornamental plants and being a general nuisance to homeowners, although it does not, in Hawaii, move into croplands. Against the unanimous recommendation of malacologists, several carnivorous land snails were introduced into Hawaii in the hope that they might control the pest. No adequate studies were made to see if, in particular, the Florida *Euglandina rosea* (Fig. 4*b*) actually *liked* to eat *Achatina* (it will only as a last resort). Instead of controlling *Achatina, Euglandina* has moved up into the native forest of the mountain slopes, where it feeds voraciously on the native *Achatinella* (see Plate 12). In some areas of Oahu *Euglandina* may be the most common land snail and seems to be drastically reducing in number the remaining native snails. Despite this bad example, authorities in Bermuda introduced *Euglandina* as a desperation means to "control" a European snail, *Otala lactea* (Fig. 4*c*). In fact, *Euglandina* will eat *Otala* only if starving. I have had young *Euglandina* die after two months on an *Otala* diet. As could have been predicted, apparently *Euglandina* is busily wiping out the native land snails of Bermuda, but only rarely disturbing the *Otala*.

Chance dispersal over millions of years and experimentations in ways of living on isolated islands have produced incredible variations in snails, insects, and plants. Unfortunately, these island organisms are in even more danger of extinction than the freshwater mollusks mentioned in Chapter VIII.

DIVERSITY

With much the same vigor that medieval philosophers expended in debating the number of angels capable of dancing on the head of a pin, modern biogeographers are exploring aspects of how many species can live in a particular area. Much of the theory and examples concerning this problem have come from studies of the island organisms. An initial observation demonstrated that there usually was a close relationship between the size of islands and the number of species living there. Subsequently the theories of island biogeography were spectacularly advanced by E. O. Wilson and Robert MacArthur. The idea that faunas have a saturation level, a maximum number of species that can live in a given area, and experimental proof of the

FIGURE 4. Land snails that have been widely introduced: (*a*) *Achatina fulica,* a 5½ inch long species originally from East Africa and carried by man throughout much of the Pacific Basin; (*b*) *Euglandina rosea,* a 2-inch-long Florida species carried by man to Hawaii, Guam, and Bermuda; (*c*) *Otala lactea,* a European edible snail usually 1½ inches in diameter, introduced into many parts of the world to serve as food.

rapidity with which both successful colonizations and extinctions can occur represent major advances in knowledge, the impact of which is still new.

Data from studies on insects, flowering plants, birds, mammals, reptiles, and ferns found on islands near continental areas confirmed the basic patterns. But data about land snails from isolated islands show an important and interesting variation. The basic difference is that on many isolated islands reached by land snails, there has been extensive local speciation. When this occurs, the number of species found on an island will greatly exceed the number predicted by the MacArthur-Wilson theory.

Pinpoints of land in the Pacific Ocean, such as Lord Howe Island (5 square miles) off Australia with 51 land snail species, and Rapa Island (south of Tahiti) with perhaps 100 species of land snails, contrast with the 58 land snail species found on Viti Levu in Fiji (4011 square miles) or the 44 species on the island of Upolu in Western Samoa (430 square miles). Biogeographic theory predicts that the larger islands should support a larger number of species. The reasons for the differences are unknown. Possibly the smaller and more isolated islands lack predators and competitors that on the large islands have reduced diversity. Possibly the smaller islands present more diversified habitats and this permitted more extensive snail evolution to occur.

But these are questions concerning the "shell makers" that await investigation. Only the briefest of introductions could be given and only the sketchiest outline of molluscan evolution and diversity could be presented herein.

The bibliography in Appendix B will serve as a guide to further information on mollusks, ranging from aids in identification to detailed anatomical and biochemical reviews. In discussing the mollusks as integrated ecological units and focusing in some of the major evolutionary changes during their history, I have tried to build an awareness of their role in the world and an appreciation of their coordinated variety. My success will be measured by your further interest and reading.

APPENDIX A

Outline Classification of Mollusks Mentioned in the Text

Class Aplacophora

- Subclass Ventroplicida
  - *Dondersia*
  - *Epimenia verrucosa*
- Subclass Caudofoveata
  - *Chaetoderma*

Class Polyplacophora

- Family Lepidopleuridae
  - *Lepidopleurus*
- Family Ischnochitonidae
  - *Ischnochiton conspicuus*
  - *Tonicella lineata*
- Family Mopaliidae
  - *Placiphorella*
- Family Chitonidae
  - *Acanthopleura granulata*
  - *Chiton tuberculatus*

Family Acanthochitonidae
Acanthochiton
Cryptoplax
Cryptochiton stelleri

Class Scaphopoda
Dentalium entalis
Dentalium indianorum

Class Monoplacophora
Cambridium
Neopilina galatheae

Class Cephalopoda
Subclass Endoceratoidea
Endoceras
Cameroceras

Subclass Actinoceratoidea
Ormoceras

Subclass Nautiloidea
Order Nautilida
Nautilus macromphalus

Order Ellesmerocerida
Plectronoceras

Subclass Ammonoidea
Pachydiscus
Hamites navarroense
Placenticeras placenta
Baculites ovatus

Subclass Coleoidea
Order Belemnitida
Pachyteuthis

Order Sepiida
Idiosepius
Sepia officinalis

Order Teuthidida
Architeuthis harveyi
Loligo
Lycoteuthis diadema
Todarodes pacificus

Order Octopoda
Cirrothauma
Argonauta argo
Hapalochlaena maculosa
Octopus hongkongensis

Class Bivalvia (classification mainly after Thiele)
? *Fordilla troyensis* (see p. 71)

Order Taxodonta
- Family Nuculidae
- Family Arcidae
  - *Arca*

Order Anisomyaria
- Family Mytilidae
  - *Fungiacava eilatensis*
  - *Lithophaga*
  - *Mytilus edulis*
- Family Isognomonidae
  - *Isognomon alata*
- Family Pectenidae
  - *Chlamys gloriosa*
  - *Pecten*
- Family Spondylidae
  - *Spondylus americanus*
- Family Limidae
  - *Lima*
- Family Ostreidae
  - *Crassostrea virginica*
  - *Gryphaea nebrascensis*

Order Modiomorphoida
- Family Actinodontidae
  - *Actinodonta*

Order Eulamellibranchia
- Suborder Schizodonta
  - Superfamily Unionacea
    - Family Margaritiferidae
      - *Margaritifera margaritifera*
    - Family Amblemidae
      - *Obovaria*
    - Family Hyriidae
      - *Velesunio ambiguus*
    - Family Unionidae
      - *Anodonta*
      - *Lampsilis siliquoidea*
      - *Leptodea fragilis*
  - Superfamily Mutelacea
    - Family Mutelidae
      - *Mutela bourguignati*
    - Family Mycetopodidae
      - *Anodontites trapesialis forbesianus*
- Suborder Heterodonta
  - Family Aetheriidae
    - *Acostaea*

Family Corbiculidae
Corbicula manilensis
Family Sphaeriidae (fingernail clams)
Family Dreissenidae
Dreissena polymorpha
Family Leptonidae
?Chlamydoconcha
Family Galeommatidae
Ephippodonta
Family Cardiidae
Corculum cardissa
Family Tridacnidae
Tridacna gigas
Tridacna squamosa
Family Veneridae
Periglypta chemnitzi
Periglypta reticulata
Mercenaria campechiensis
Pitar dione
Family Psammobiidae
Ascitellina urinatoria
Family Semelidae
Scrobicularia plana
Family Tellinidae
Macoma nasuta
Family Mactridae
Spisula
Family Solenidae
Ensis
Cultellus pellucidus
Solen
Family Hiatellidae
Panope generosa
Family Myacidae
Mya arenaria
Family Pholadidae
Barnea costata
Family Teredinidae
Family Clavagellidae
Brechites
Clavagella

Order Septibranchia

Family Poromyacidae
Poromya
Cetoconcha

Family Cuspidariidae
Cuspidaria

Class Gastropoda (after Fretter and Graham, Solem)
Subclass Prosobranchia
Order Bellerophontina
Family Helcionellidae
Helcionella
Family Coreospiridae
Coreospira
Order Diotocardia (= Archaeogastropoda)
Family Pleurotomariidae
Perotrochus
Family Haliotidae
Haliotis asinina
Family Scissurellidae
Scissurella crispata
Family Fissurellidae
Emarginula
Fissurella
Family Patellidae
Patella
Family Acmaeidae
Family Trochidae
Family Turbinidae
Astraea phoebia
Guildfordia triumphans
Turbo
Family Neritidae
Nerita polita
Family Helicinidae
Dawsonella meeki
Helicina
Sturanyella plicatilis

Order Monotocardia
Suborder Taenioglossa (= Mesogastropoda)
Superfamily Cyclophoracea
Family Cyclophoridae
Cyclophorus
Pterocyclos prestoni
Rhiostoma hainesi
Family Poteriidae
Aperostoma mexicanum palmeri
Ostodes tiara
Family Pupinidae

Pupinella simplex
Rhaphaulus lorraini
Family Diplommatinidae
Opisthostoma coronatum
Opisthostoma retrovertens
Family Viviparidae
Cipangopaludina
Family Ampullariidae
Pomacea cumingii
Pomacea paludosa
Superfamily Valvatacea
Family Valvatidae
Valvata
Superfamily Littorinacea
Family Littorinidae
Littorina littorea
Littorina neritoides
Littorina obtusata
Littorina saxatilis
Littorina ziczac
Family Pomatiasidae
Colobostylus
Opisthosiphon bahamensis
Pomatias elegans
Superfamily Rissoacea
Family Hydrobiidae
Amnicola
Gocea ohridiana
Family Truncatellidae
Truncatella
Family Homalogyridae
Ammonicera japonica
Superfamily Cerithiacea
Family Turritellidae
Vermicularia
Family "Thiaridae"
Pleurocera
Superfamily Strombacea
Family Strombidae
Lambis crocata
Lambis lambis
Superfamily Calyptraeacea
Family Calyptraeidae

Superfamily Cypraeacea
- Family Amphiperatidae
  - *Calpurnus verrucosus*
- Family Cypraeidae
  - *Cypraea cribraria*
  - *Cypraea limacina facifer*
  - *Cypraea saulae nugata*
  - *Cypraea subviridis*

Superfamily Naticacea
- Family Naticidae

Superfamily Epitoniacea (= Ptenoglossa)
- Family Janthinidae
  - *Janthina*

Superfamily Aglossa
- Family Eulimidae
- Family Entoconchidae

Superfamily Heteropoda

Superfamily Tonnacea
- Family Cassidae
- Family Cymatiidae
  - *Biplex perca*
  - *Charonia tritonis*
  - *Cymatium rubeculum*
- Family Bursidae
- Family Tonnidae
- Family Ficidae
  - *Ficus subintermedius*

Suborder Stenoglossa (= Neogastropoda)

Superfamily Muricacea (family limits uncertain)
- *Drupa ricinus*
- *Drupina grossularia*
- *Murex acanthostephes*
- *Murex (Bolinus) cornutus*
- *Murex pecten*
- *Typhina pavlova*

Superfamily Buccinacea
- Family Buccinidae
  - *Clea*
- Family Melongenidae
  - *Busycon*
  - *Melongena corona*
  - *Syrinx aruanus*
- Family Nassariidae
- Family Fasciolariidae
  - *Fusinus*

Opeatostoma pseudodon (= *Leucozonia cingulifera*)

Superfamily Volutacea

Family Olividae

Family Mitridae

Family Volutidae

Amoria canaliculata

Amoria maculata

Cymbiolacca pulchra

Melo amphora

Family Marginellidae

Marginella

Rivomarginella

Superfamily Conacea (= Toxoglossa)

Family Conidae

Conus ammiralis

Conus geographus

Conus marmoreus

Conus striatus

Conus textile

Family Terebridae

Terebra areolata

Family "Turridae"

Subclass Opisthobranchia

Order Pyramidellacea

Family Pyramidellidae

Order Gymnosomata (shelless pteropods)

Order Sacoglossa

Family Juliidae

Berthelinia limax

Order Nudibranchia

Superfamily Doridacea

Family Hexabranchidae

Hexabranchus

Family Doridae

Family Phyllidiidae

Phyllidia varicosa

Order Acochlidiacea

Subclass Pulmonata

Superorder Systellommatophora

Order Onchidiacea

Family Onchidiidae

Order Soleolifera
Family Veronicellidae
Veronicella
Family Rathouisiidae
Superorder Basommatophora
Family Ellobiidae
Pythia pachyodon
Family Lymnaeidae
Lymnaea stagnalis
Superorder Stylommatophora
Family position uncertain
Anthracopupa ohioensis
Maturipupa
Dendropupa vetusta
"Zonites" priscus
Order Orthurethra
Family Achatinellidae
Achatinella bellula
Achatinella juddii
Achatinella mustelina
Family Amastridae
Family Pupillidae
Gyliotrachela depressispira
Family Partulidae
Samoana canalis
Order Mesurethra
Family Clausiliidae
Family Strophocheilidae
Strophocheilus
Order Sigmurethra
Suborder Holopodopes
Family Achatinidae
Achatina fulica
Family Streptaxidae
Discartemon hypocrites
Oophana
Family Haplotrematidae
Family Rhytididae
Paryphanta busbyi
Ptychorhytida aulacospira
Family Aperidae
Apera
Family Bulimulidae
Liguus fasciatus solidus form *pictus*
Orthotomium pallidior

Suborder Aulacopoda
Superfamily Arionacea
Family Endodontidae
Thaumatodon spirrhymatum
Family Charopidae
Flammulina
Paryphantopsis
Pilsbrycharopa
Ptychodon microundulata
Ranfurlya
Suteria ide
Maoriconcha oconnori
Family Arionidae
Geomalacus
Milax gagates
Prophysaon
Arion
Family Philomycidae
Superfamily Succineacea
Family Succineidae
Succinea norfolkensis
Family Athoracophoridae
Pseudaneitea
Superfamily Limacacea
Family Helicarionidae
Austenia
Cryptaustenia
Diastole conula
Durgella
Fanulum testudo
Girasia
Helicarion
Megaustenia
Muangnua
Parvatella
Family Urocyclidae
Family Aillyidae
Family Zonitidae
Velifera
Family Trochomorphidae
Trochomorpha
Family Limacidae
Limax cinereoniger
Family Parmacellidae

Family Testacellidae
Testacella

Suborder Holopoda
Superfamily Polygyracea
Family Polygyridae
Ashmunella kochi ambyla
Stenotrema barbatum

Superfamily Oleacinacea
Family Oleacinidae
Euglandina rosea

Superfamily Camaenacea
Family Camaenidae
Amphidromus
Globorhagada
Labyrinthus otis
Meridolum gulosum
Papuina boivinii
Papuina phaeostoma
Pleuroxia
Polydontes
Rhagada

Family Bradybaenidae
Helicostyla

Family Helminthoglyptidae
Leptarionta trigoniostoma
Micrarionta veitchii
Polymita
Sonorella dalli

Family Helicidae
Arianta arbustorum
Eremina desertorum
Helix aperta
Helix aspersa
Helix pomatia
Otala lactea

APPENDIX B
Selected References for Further Reading

Each year several hundred scientists publish several thousand articles about mollusks in technical journals. These publications range from short notes on natural history observations to huge monographs on a genus or family and massive reports on mollusks from a geographic area. More than 95,000 such articles have been published to date, with more being added each year.

Some of this literature attempts to synthesize knowledge and to communicate it to an audience broader than other professional scientists. The following brief list attempts to provide sources that will answer additional questions without overwhelming the reader with technicalities and which will in turn lead to the more specialized literature.

Some of these references are out of print, but are available in bigger libraries. As a guide to users, the books are keyed as to price range: under \$5.00 (I for inexpensive), \$5.00 to \$15.00 (M for moderate cost), and over \$15.00 (E for expensive). The list is organized by the types of questions answered.

GENERAL BOOKS ON INVERTEBRATES

Ralph Buchsbaum. 1958. *Animals without backbones.* University of Chicago Press. A well illustrated textbook. (M)

Ralph Buchsbaum & Lorus J. Milne. 1960. *The lower animals.* Doubleday, New York. A more popular account with color illustrations. (M)

Alister Hardy. 1956. *The open sea, the world of plankton.* Collins, London. A modern classic. (M)

Robert W. Pennak. 1953. *Fresh-water invertebrates of the United States.* Ronald Press, New York. Excellent account of their natural history and keys to many groups. (E)

David Nichols. 1962. *Echinoderms.* Hutchinson University Library, London. Part of an excellent series, all highly recommended. (I)

E. J. W. Barrington. 1965. *The biology of Hemichordata and Protochordata.* Oliver & Boyd, London. Very good. (I)

L. H. Hyman. 1939–1967. *The invertebrates.* Vols. 1–6. McGraw-Hill, New York. The most important summary volumes in this century, but highly technical and very expensive.

GENERAL BOOKS ON MOLLUSKS

John Morton. 1967. *Molluscs.* Hutchinson University Library, London. Best general survey organized according to life function. Requires biological background. (I)

W. D. Russell-Hunter. 1968. *A biology of the lower invertebrates.* MacMillan, New York. (I)

A. H. Cooke. 1895. *Mollusca.* Cambridge Natural History Series. Vol. 3. Recently reprinted and still excellent. (M)

Paul Pelseneer. 1906. Vol. 5 in *A treatise on zoology,* edited by E. R. Lankester. Reprint by Asher & Co., Amsterdam in 1964. Excellent survey of anatomy by a leading scientist. (M)

K. M. Wilbur & C. M. Yonge. 1964 and 1966. *The physiology of Mollusca.* Academic Press, New York. Standard symposia volumes providing copious references. (E)

L. H. Hyman. 1967. *The invertebrates.* Vol. 6. *Mollusca I.* McGraw-Hill, New York. Covering Aplacophora, Monoplacophora, Polyplacophora, and Gastropoda. A classic. (E)

CLASSIFICATION OF MOLLUSKS

Johannes Thiele. 1929–1935. *Handbuch der systematischen Weichtierkunde.* Four parts. Gustav Fischer, Jena. Has been reprinted. The only account of all living mollusks down to subgenus level. In German. A classic. (E)

W. Wenz & A. Zilch. 1938–1960. *Handbuch der Paläozoologie: Gastropoda.* Gebrüder Borntraeger, Berlin. Illustrations and short diagnoses for each genus and subgenus of both living and fossil snails. Superb, but costs several hundred dollars.

Raymond C. Moore, editor. 1959– . *Treatise on invertebrate paleontology.* Parts I through N. Geological Society of America, New York. Published parts on Bivalvia and Cephalopoda excellent for general data plus cryptic synopsis of classification. A continuing series. (M to E)

CEPHALOPODS

Frank W. Lane. 1957. *Kingdom of the octopus.* Jarrolds, London. A comprehensive review by a skilled popular writer. Extensive bibliography. (M)

A. Packard. 1972. *Cephalopods and fish: the limits of convergence.* Biological reviews of the Cambridge Philosophical Society. Vol. 47, No. 2, pp. 241–307. Excellent and very thought-provoking review with extensive bibliography. (M)

M. J. Wells. 1962. *Brain and behaviour in cephalopods.* Heinemann, London. A modern classic on behavior and learning. (I)

CLAMS

C. M. Yonge. 1960. *Oysters.* Collins, London. Excellent general account. (M)

Steven M. Stanley. 1970. *Relation of shell form to life habits of the Bivalvia.* Memoir 125, Geological Society of America, New York. A highly significant technical report. (M)

R. D. Purchon. 1968. *The biology of the Mollusca.* Pergamon Press, Oxford. Bivalve section better than others. (E)

Fritz Haas. 1929–1956. Bivalvia. In Bronn, *Klassen und Ordnungen des Tierreichs.* Vol. III. Monumental several-volume survey. Very expensive.

Also see *Treatise on invertebrate paleontology,* listed above.

SNAILS

Vera Fretter & A. Graham. 1962. *British prosobranch molluscs.* Ray Society, London. A monumental survey of biology, functional anatomy, and evolution of the prosobranchs. Best single book on mollusks in this century. (E)

L. H. Hyman. *The invertebrates.* Vol. 6. *Mollusca I.* Massive synthesis of data and huge bibliography. (E)

IDENTIFICATION GUIDES

Such books generally cover either a geographic *area* or survey *one* particular genus or family of mollusks. A few are published in journal form over a number of years, covering the mollusks of a large geographic area family by family.

NORTH AMERICAN MARINE MOLLUSKS

R. Tucker Abbott. 1968. *Seashells of North America.* Golden Press, New York. A best buy with superb illustrations and packed with facts. Covers only slightly less than his much more expensive *American seashells* (Van Nostrand, 1955). (I)

E. L. Bousfield. 1960. *Canadian Atlantic shells.* National Museum of Canada, Ottawa. Excellent regional guide. (I)

Johnsonia. 1941– . *Monographs of the marine mollusks of the Western Atlantic.* Department of Mollusks, Museum of Comparative Zoology, Harvard University. Irregularly issued studies of families or genera. Comprehensive and good, now in Vol. 5. (E)

Morris, Percy A. 1973. *A field guide to shells of the Atlantic and Gulf coasts and the West Indies.* Third edition edited by W. J. Clench. Houghton Mifflin, Boston. An excellent revision of a popular book. (M)

Louise M. Perry & Jeanne S. Schwengel. 1955. *Marine shells of the western coast of Florida.* Paleontological Research Institute, Ithaca, New York. Superb plates and good text. (M)

Jean Andrews. 1971. *Seashells of the Texas coast.* University of Texas Press, Austin. Best on area. (E)

G. L. Warmke & R. T. Abbott. 1961. *Caribbean seashells.* Livingston, Narbeth, Pennsylvania. Good for identifications. (M)

Tom Rice. 1971. *Marine shells of the Pacific Northwest.* Ellison Industries, Edmonds, Washington. Handy field guide with color plates. (I)

James H. McLean. 1969. *Marine shells of Southern California.* Los Angeles County Museum of Natural History, Los Angeles. Well-done handbook with black and white photographs. (I)

A. Myra Keen. 1971. *Seashells of tropical West America.* Stanford University Press, Stanford. A monumental survey of all marine mollusks from Panama to San Diego. (E)

EUROPEAN MARINE MOLLUSKS

A. Graham. 1971. *British prosobranchs.* Synopses of the British Fauna (New Series) No. 2. Academic Press, London. Very good for identifications. (I)

Norman Tebble. 1966. *British bivalve seashells.* British Museum (Natural History), London. Very good manual, with the paperback edition a best buy. (I)

INDO-PACIFIC MARINE MOLLUSKS

Tadashige Habe. 1971. *Shells of Japan.* 139 pp., many color plates. Hoikusha Publishing Company, Ltd., Osaka. A superbly fact-packed book and a best buy. (I)

B. R. Wilson & K. Gillett. 1971. *Australian shells.* Charles Tuttle, Rutland, Vermont.

Mainly tropical gastropods, but unique for the large number of living mollusks portrayed. (E)

A. W. B. Powell. 1957. *Shells of New Zealand.* Whitcombe & Tombs, Auckland. Check list and many plates. (M)

J. H. Macpherson & C. J. Gabriel. 1962. *Marine molluscs of Victoria (Australia).* National Museum of Victoria, Melbourne. Excellent account for Southern Australia. (M)

K. H. Barnard. 1953. *A beginners guide to South African shells.* Maskew Miller Ltd., Cape Town. A good field guide. (M)

Indo-Pacific Mollusca. 1959– . Equivalent of *Johnsonia* and issued at irregular intervals. Some color plates. Department of Mollusks, Delaware Museum of Natural History, Greenville. (E)

NOTE: Numerous books in English have been published in Japan with lavish color plates. Most are excellent (M to E). Several other Australian and Pacific Ocean books also are quite good, but those given above provide the best introduction to this vast fauna.

FRESHWATER MOLLUSKS

F. C. Baker. 1928. *The freshwater Mollusca of Wisconsin.* Parts I (*Gastropoda*) & II (*Pelecypoda*). Bulletin 70, Wisconsin Geological and Natural History Survey. While many workers do not accept his ideas as to what is a species, this is the best general introduction in English. Recently reprinted. (E)

B. Walker. 1918. *A synopsis of the classification of the freshwater Mollusca of North America.* University of Michigan Museum of Zoology, Ann Arbor, Michigan. Check list and discussion. (I)

Harold D. Murray & A. B. Leonard. 1962. *Handbook of Unionid mussels in Kansas.* Museum of Natural History, University of Kansas, Lawrence, Kansas. Excellent survey. (I)

Paul W. Parmalee. 1967. *The freshwater mussels of Illinois.* Illinois State Museum Society, Illinois State Museum, Springfield, Illinois. Plates not as good as the previous book, but a useful manual. (I)

William C. Starrett. 1971. *A survey of the mussels of the Illinois River, a polluted stream.* Illinois Natural History Survey Bulletin, Urbana, Illinois. Sobering account of how the mussel fauna of one river has been almost exterminated by pollution. Several excellent color plates. (I)

LAND MOLLUSKS

Henry A. Pilsbry. 1939–1948. *Land Mollusca of North America (north of Mexico).* Monograph No. 3, Academy of Natural Sciences of Philadelphia, Pennsylvania. Basic reference work in four parts that will not be replaced in this century. (E)

John B. Burch. 1962. *How to know the eastern land snails.* Wm. C. Brown, Dubuque, Iowa. Handy field guide with keys and line illustrations. (I)

Arthur E. Ellis. 1926. *British snails.* Clarendon Press, Oxford. Excellent introduction and identification guide, recently reprinted. (M)

Frank C. Baker. 1939. *Fieldbook of Illinois land snails.* Illinois Natural History Survey Manual 2. OUT OF PRINT. Excellent guide occasionally available on book lists of natural history book dealers. (I to M)

MONOGRAPHS OF MARINE FAMILIES POPULAR WITH COLLECTORS

C. M. Burgess. 1970. *The living cowries.* A. S. Barnes & Co., New York. Good plates, but text with some errors. (E)

C. S. Weaver & John E. du Pont. 1970. *Living volutes. A monograph of the recent Volutidae of the world.* Delaware Museum of Natural History, Greenville, Delaware. Good. (E)

R. F. Zeigler. 1969. *Olive shells of the world.* Ziegler & Porreca, West Henrietta, New York. Nice color plates. (M)

BOOKS ON SHELL COLLECTING

How to collect shells. American Malacological Union. Order from Marian Hubbard, 3957 Marlow Court, Seaford, New York. Varied articles on collecting and studying shells. Several editions have appeared. (I)

Kathleen Y. Johnstone. 1957. *Sea treasure.* Houghton Mifflin Company, Boston. Excellent for the beginning collector. (M)

R. T. Abbott. 1962. *Sea shells of the world.* Golden Nature Guide. Golden Press, New York. Superb paperback with numerous color plates. A best buy. (I)

S. Peter Dance. 1966. *Shell collecting, an illustrated history.* Faber and Faber, London. Fine historical account of early collectors and famous workers. (M)

R. J. L. Wagner & R. T. Abbott. 1967. *Van Nostrand's standard catalog of shells.* Van Nostrand-Reinhold, Princeton, New Jersey. List of names for many families and price ranges, some illustrations, very well done. (M)

ORGANIZATIONS AND JOURNALS

Two organizations in the United States combine scientists, amateur collectors, university students, and interested laymen. Both hold annual meetings at which symposia and lectures are presented.

American Malacological Union, a national group established in the early 1930s. Contact Mr. Paul Jennewein, Box 394, Wrightsville Beach, North Carolina 28480.

Western Society of Malacologists, a primarily western group established in 1968. Contact Ralph Fox, California Academy of Sciences, Golden Gate Park, San Francisco, California 94118.

Many regularly issued journals publish only original research articles on mollusks. The following are among the largest and most important journals:

Archiv für Molluskenkunde (Dr. A. Zilch, Senckenbergische Naturforschende Gesellschaft, Senckenberg-Anlage 25, 6 Frankfurt-am-Main 1, West Germany).

Basteria (Dr. H. E. Coomans, Zoölogisch Museum, Plantage Middenlaan 53, Amsterdam-C, Netherlands).

Journal of Conchology (Mrs. E. B. Rands, 51 Wynchwood Avenue, Luton, Bedfordshire, United Kingdom).

Journal of the Malacological Society of Australia (Dr. Brian Smith, National Museum of Victoria, Russell Street, Melbourne, Victoria, Australia 3000).

Malacologia (c/o Department of Mollusks, Academy of Natural Sciences, 19th & the Parkway, Philadelphia, Pennsylvania 19103).

Nautilus (Mrs. H. B. Baker, 11 Chelten Road, Havertown, Pennsylvania 19083).

Proceedings of the Malacological Society of London (Dr. Joyce E. Rigby, Department of Biology, Queen Elizabeth College, Campden Hill Road, London, W 8, United Kingdom).

Venus (Malacological Society of Japan, c/o National Science Museum, Ueno Park, Taito-ku, Tokyo, Japan).

Veliger (Mrs. Jean R. Cate, P. O. Drawer R, Sanibel, Florida 33957).

APPENDIX C
Glossary of Terms

This glossary lists technical terms unique to mollusks and words whose commonly used meanings are different from the restricted ones used in the text.

ACELLULAR Members of the kingdom Protista whose bodies consist of one unit or a colony of units.

ACTINODONT Clam hinge teeth that radiate from the umbos.

ADAPTIVE RADIATION When a group of organisms diversifies into a number of different ways of living.

ADDUCTOR MUSCLES Used to close the valves of a bivalved shell.

AESTIVATE Reduced body metabolism to last over a dry spell; the wet-dry climate equivalent of hibernation.

ALBUMEN GLAND Provides nutritive material for fertilized egg.

AMMONOID Extinct cephalopod that dominated Mesozoic seas.

ANAL SLIT Notch or slit in archaeogastropod snail shell through which body wastes and exhalent water currents pass.

ANIMAL A multicellular organism that consumes its food (*see* CONSUMER).

ANISOMYARIAN Clams with adductor muscles that are unequal in size.

ANUS External opening through which undigested food passes.

APERTURE Opening of snail shell.

APICAL WHORLS Part of shell formed in egg or right after larva changes from swimming veliger; first formed shell.

APLACOPHORA Solenogasters; a class of mollusks.

ARCHAEOZOIC ERA Part of prefossil record of geologic time; early portion of 600 to 4500 million years before the present.

ARIONACEAN Member of the land snail Superfamily Arionacea.

ARTICULAMENTUM Inner layers of a chiton shell.

AULACOPOD Pulmonate land snail with longitudinal grooves along each side of the foot.

AUSTRALIAN Biogeographic region consisting of New Guinea, Australia, New Zealand, and adjacent islands, sometimes Pacific Islands included; adj., an organism native to this region.

BARRIER Any projection or growth that narrows the aperture of the snail shell.

BASAL PLATE Portion of radular tooth attached to radular membrane.

BELEMNITE Extinct group of cephalopods that were ancestors of the modern squids.

BELLEROPHONTS Earliest known snails with a spiral shell.

BIOGEOGRAPHY Study of living organism distribution.

BIVALVIA The class of mollusks including clams, oysters, and mussels.

BODY WHORL Last coiling of a snail shell before the aperture.

BUCCAL Refers to the radula, mouth cavity, and mass of muscles that operate the radular apparatus.

BYSSUS Threads secreted by the foot of a clam to anchor it to some other object.

CAPTACULA Fine food-catching tentacles of scaphopods (tooth shells).

CARDINAL TEETH Hinge teeth of clams that are near the umbos.

CARNIVORE An animal that eats other animals.

CENOZOIC ERA The last 63 million years of geologic time.

CENTRAL TOOTH Single tooth in the middle of a snail's radula.

CEPHALOPODA The class of mollusks including squids, octopuses, and cuttlefish.

CERATA Fingerlike projections on the back of sea slugs, often used in respiration or for protection.

CHEMORECEPTORS Sensors that detect chemicals, as in our smell and taste.

CHITON Member of the Class Polyplacophora; coat-of-mail shells.

CIRRUS Retractile tip of the tentacles in the living nautiloid, *Nautilus* (chambered nautilus).

COLEOID Ordinal name for squids, octopuses, and cuttlefish.

COLUMELLAR That portion of the snail shell nearest to the shell axis; lower inner portion of apertural lip.

COLUMELLAR MUSCLE Upper end of free muscle system that attaches a snail to its shell.

COMMENSALISM Different species that live in close association, but without much effect on each other.

CONSUMER An organism that obtains organic matter by ingesting (eating) other organisms (*see* PRODUCER, REDUCER); may be an animal or a protozoan.

CONTINENTAL ISLANDS Islands that had dry land connections with a continental mass at one time to another.

CUSPS Sharp elevated cutting edges of a radular tooth.

DART APPARATUS Accessory genital structure used in mating as an aid in species recognition.

DEUTEROSTOMIA Group of animal phyla leading to and including the Phylum Chordata (*see* PROTOSTOMIA).

DIMYARIAN A clam with two adductor muscles.

DIOTOCARDIA Group of prosobranch snails with (primitively) two sets of pallial organs; also called Archaeogastropoda.

DIPLEURULA Larval form characteristic of the Deuterostomia.

DISJUNCTIVE DISTRIBUTION Species or group of species found in two or more areas that are separated by a significant geographic gap.

DORSAL Top or upper side.

EDENTULOUS Condition in which the hinge teeth of a clam are greatly reduced or completely absent.

EPIFAUNA Term for organisms that are attached or move on the surface of the sea or freshwater bottom sediments (*see* INFAUNA).

EPIPHRAGM Sheet(s) of mucus or calcified mucus secreted by a pulmonate mollusk to seal the shell aperture.

EPIPODIUM Sensory fringe of lobes or tentacles around the body of some archaeogastropods.

ESCUTCHEON Part of clam shell next to an external ligament.

ETHIOPIAN Biogeographic region including Africa south of the Sahara and adjacent islands; adj., an organism native to this region.

EXHALANT SIPHON The siphon used by a clam to expel water from its body.

FOSSIL Any remains or direct evidence of an organism that has been preserved in a rock.

FUNGAE Multicellular organisms that release digestive enzymes from their body, then adsorb the partly digested food; fungi, molds, mushrooms, rusts; reducer organisms (*see* CONSUMER, PRODUCER).

GASTROPODA The class of mollusks for snails and slugs.

GENITAL PORE External opening(s) of the genitalia.

GILLS Organs of respiration in the mantle cavity (usually) of most water dwelling mollusks.

GIRDLE Flexible outer portion of chiton in which the shell plates are embedded.

GLOCHIDIUM Larval clam found in the freshwater Unionacea that is parasitic on fishes.

HAUSTORIUM Larval clam found in the African Mutelidae, parasitic on freshwater fishes.

HELICOID General term for a low to medium asymmetrically coiled shell of a land snail; sometimes refers to Superfamilies Camaenacea, Polygyracea, Helicacea.

HERBIVORE An animal or protozoan that feeds on producer organisms (plants or algae) (*see* CARNIVORE, REDUCER, PRODUCER).

HERMAPHRODITIC Individuals that produce both sperm and eggs.

HERMAPHRODITIC DUCT Tube in pulmonate and opisthobranch snails from ovotestis to area where fertilization takes place; used by sperm and eggs.

HETERODONT Clam hinge teeth that are differentiated into cardinal and laterals.

HINDGUT Section of digestive system running forward to anus.

HINGE TEETH Opposing pits and projections along dorsal margin of clam shell that interlock and prevent valves from moving backward or forward in relation to each other.

HOLOPOD Pulmonate land snail with no longitudinal grooves along each side of the foot.

HOOD Dorsal shield in *Nautilus* that acts as a closure to the shell aperture when the animal is retracted.

HYOLITHA A problematic group of fossils; may not be mollusks.

HOLARCTIC Term used when Palearctic and Nearctic Regions are combined into a single biogeographic region; adj., an organism native to this area.

HYPOBRANCHIAL GLAND Mucus gland that aids in cleaning mantle cavity of marine snails.

HYPONOME The funnel or jet propulsion device of cephalopods.

INFAUNA Organisms that burrow in sea bottom or freshwater sediments (*see* EPIFAUNA).

INHALENT SIPHON Clam siphon used to bring water into the body.

INSERTION TEETH Parts of chiton plates that unite them to the girdle.

ISODONT Type of hinge teeth in *Chlamys* and *Spondylus*.

ISOMYARIAN Clam with two adductor muscles that are equal in size.

JUGAL, JUGAL SINUS Parts of a chiton plate.

LAND BRIDGE A narrow strip of land connecting two larger land areas, such as Panama joining Central and South America.

LASIDIUM Clam larvae found in the South American Mycetopodidae.

LATERAL TEETH Clam: hinge teeth removed from area of umbos; radula: tooth type nearest the central tooth, often used in cutting.

LIGAMENT Elastic band that unites the two calcified parts of a bivalved shell along (usually) dorsal margin.

LIMACACEAN Land snail member of the superfamily Limacacea.

LIMITS OF TOLERANCE Extremes of conditions of an environmental factor between which an organism can survive.

LIMPET Any snail that has a conical shell, normally living on rocks or plants in the tidal zone.

LUNULE Area on clam valve just anterior of umbos.

MANTLE Fold from body wall that secretes the shell (when present).

MANTLE CAVITY Space formed by extending the mantle out and down into which anus, excretory, and reproductive pores open; normally containing gills and chemoreceptors.

MARGINAL TEETH Teeth on each side of a radula, usually of a different shape from the laterals and used to collect food particles torn loose by the central and/or laterals.

MATTHEVA A problematic group of fossils.

MESOZOIC ERA 63 to 230 million years ago.

MESURETHRA An order of pulmonate land snails.

METAPODIUM Posterior lobe of a snail's foot; in conchs it ends in the operculum.

MONERA Kingdom containing blue-green algae and bacteria; organisms that lack a separated nucleus.

MONOMYARIAN A clam with only one adductor muscle.

MONOPLACOPHORA A class of partly segmented mollusks.

MONOTOCARDIA An order of prosobranch snails in which there is only one set of pallial organs.

MUCUS Slippery secretion from the skin or glands of mollusks.

MUTUALISM When different species that live in close association both benefit from the association.

NACRE The pearly or iridescent layers of some shells.

NAUTILOID Term applied to the mostly Paleozoic cephalopods.

NEARCTIC Biogeographic region including North America and nontropical Mexico; adj., an organism native to this region.

NEOTROPICAL Biogeographic region including tropical Mexico, Central America, South America, and West Indies; adj., an organism native to this region.

NICHE The role of an organism in the area in which it lives.

OCEANIC ISLANDS Islands that have never been connected to any continent.

ODONTOPHORE General term for the radula and its supporting structures.

OPERCULUM Horny or calcareous disk on the foot of a snail that often can seal the shell opening when the snail has retracted.

OPISTHOBRANCHIA A subclass of snails.

ORIENTAL Biogeographic region including India, South China, Southeast Asia, Philippine Islands, and Indonesia east to Celebes; adj., an organism native to this region.

ORTHURETHRA An order of pulmonate land snails.

OSPHRADIUM A chemoreceptor found in the mantle cavity of many mollusks.

OVOTESTIS Organ producing sperms and eggs in hermaphroditic (which see) mollusks.

OVIPARITY Eggs are laid and young develop inside the egg after it has been laid.

OVOVIVIPARITY Eggs enclosed by a membrane or shell and held inside the body until ready to hatch.

PALEARCTIC Biogeographic region including Europe, Asia north of the Himalayas, and Africa north of the Sahara; adj., an organism native to this region.

PALEOZIC ERA 230 to 600 million years ago.

PALATAL Outer wall of aperture in a snail shell.

PALLIAL CAVITY A mantle cavity with a narrowed opening.

PALLIAL LINE Point at which mantle muscles attach to a clam shell.

PALLIAL SINUS Area in which siphonal muscles attach to a clam shell.

PAPILLAE Short to long protrusions or warts on soft tissues, as on the mantle of *Cypraea*.

PARAPODIA "Legs" of annelid worms; lateral body projections in some opisthobranch mollusks.

PARIETAL Inner wall of aperture in a snail shell.

PEARL A foreign object covered with layers of shell by a clam.

PERIOSTRACUM Outer organic layer(s) of a shell.

PHOTOSYNTHESIS Process by which light energy is used in the presence of chlorophyll to manufacture organic chemicals from water and carbon dioxide, with oxygen given off as a waste product.

PLANKTON Organisms that float or very weakly swim in the upper waters of the ocean.

PLANT A multicellular producer organism (*see* CONSUMER, REDUCER).

PLATE One of the shell parts of a chiton: anterior or head; posterior or tail; intermediate.

PLEURAL Part of the plate in a chiton.

PNEUMOSTOME Circular opening to the pallial cavity in a pulmonate snail.

POLYPLACOPHORA A class of mollusks for the chitons.

PROBOSCIS The protruded feeding organ of a snail.

PRODUCER An organism that manufactures organic chemicals from inorganic matter (*see* CONSUMER, REDUCER).

PROPODIUM Anterior portion of a snail's foot.

PROSBRANCHIA A subclass of snails.

PROTEROZOIC Part of the prefossil record, the later part of 600 to 4500 million years ago.

PROTISTA Kingdom of acellular organisms, including protozoans and simple algáe.

PROTOSTOMIA Group of animal phyla leading to the Arthropoda.

PULMONATA A subclass of snails.

RACHIGLOSSATE RADULA One with two laterals and a single central tooth.

RADULA The "toothed tongue" or feeding organ of most mollusks.

RADULAR CARTILAGE Supporting structure of radula.

RADULAR MEMBRANE Surface on which teeth are fastened that grows forward as worn teeth are discarded.

REDUCER An organism that obtains energy by breaking down organic matter into simpler chemicals (*see* PRODUCER, CONSUMER).

RESILIUM Internal ligament of a clam with isodont or modified heterodont dentition.

RETRACTOR MUSCLES Muscles that pull back part of the body.

RHIPIDOGLOSSATE RADULA A radula with several laterals and very many tiny marginal teeth.

ROSTROCONCHIA A group of fossil clams with small spiral shell remnant showing dorsal valve fusion.

SACOGLOSSAN Opisthobranchs specialized to feed on algae.

SCAPHOPODA A class of mollusks including tusk or tooth shells.

SEPTA Internal partitions in shelled cephalopods.

SHELL Protective (usually) covering of calcium and an outer organic layer secreted by the mantle.

SHELL AXIS Imaginary line around which a spiral shell coils.

SIGMURETHRA An order of pulmonate land snails.

SIPHON A tube, usually formed from the mantle, through which a current of water passes.

SIPHONAL CANAL An extension of the snail shell to protect a long siphon.

SIPHUNCLE Tube in shelled cephalopods connecting the animal and the chambers of the shell.

SLIT RAYS Paths for sensory organelles to pass through a chiton plate.

SLUG Any snail with a greatly reduced or no shell.

SNAIL Member of the Class Gastropoda.

SOLENOGASTER Member of the Class Aplacophora.

SPERMATHECA Sperm storage organ.

SPERMATOPHORE Soft or hard capsule in which many sperms are transferred from one individual to another.

SPIRE Part of snail shell above body whorl (which see).

SPIRE ANGLE The angle at which the spire widens.

STENOTHECOIDA A group of problematic fossils.

SUTURAL LAMINAE Part of a chiton plate.

SUTURE Line where two whorls of a snail shell overlap.

SYMBIOSIS Different species, that live in close association (see commensalism, mutualism).

SYSTELLOMMATOPHORA A Superorder of the Pulmonata.

TAENIOGLOSSATE RADULA A radula with two marginals and a lateral tooth on each side of the central tooth.

TAIL FAN Retractor muscles that pull the tail of a snail into its shell.

TAXODONT Clams with very many hinge teeth.

TEGMENTUM Outer layers of a chiton shell.

TORSION Process during which the visceral hump and mantle cavity of a larval snail rotate 180° counterclockwise within a few hours.

TOXOGLOSSAN Advanced prosobranch snails with a poison gland.

TROCHOPHORE Larval type characteristic of protostome animals.

UMBILICUS Basal opening in a snail shell.

URETER Closed tube from kidney to the excretory pore.

UMBOS First formed section of a clam valve.

VALVES Calcified parts of a clam shell.

VARIX Resting phase during snail growth marked by a rib or thickening of the shell.

VENTRAL Lower or bottom surface.

VISCERAL HUMP Section of mollusk never extended from the shell (when shell is present).

VIVIPARITY Young nourished directly by parent throughout development inside body of parent, then born alive.

APPENDIX D
Care and Feeding of Mollusks

Many mollusks have extremely precise needs in regard to temperature, humidity, water salinity, food, or other conditions in their environment, and are thus extremely difficult to maintain under laboratory conditions for study and experimentation. Other species of mollusks can withstand a wide range of variation or adapt easily to life in small containers at household temperatures. Only trial and error will reveal whether a particular species can be kept at home successfully. Since more than 1000 species of marine mollusks live in the seas off Florida, most mainland states have more than 100 land snail species within their borders, and ponds and streams of Eastern North America, for example, have a total of more than 2000 freshwater species, only a few general suggestions can be given here. It is impossible to consider individual species.

There are a few principles to follow and some specific hints that relate to mollusks. A good starting place is to follow the basic recommendations in books on home aquariums and terrariums—that is, provide a sample of the environment in which the animal usually lives. Aquatic organisms must get sufficient oxygen circulated through the water, which is easily provided by an aquarium pump aerator. All organisms must have food, but only in quantities that can be consumed and not so much that uneaten food will accumulate, then decay, and foul the container. Minor quantities of excess food and waste products from the animal can be taken care of by naturally occurring reducer organisms, but, particularly with land snails, cleaning of their home at biweekly intervals increases the chance of success.

In addition to the general advice given in an aquarium or terrarium guide, a guide to collecting mollusks, such as the American Malacological Union's *How to Collect Shells,* will help you to find live mollusks and suggest the basic equipment needed to keep them alive. Identification guides will tell you the name of the species, and the general nature of its food may have been mentioned in earlier pages of this book. The following tips are extracted from articles by and conversations with both amateur and professional malacologists, in addition to personal experience.

MARINE MOLLUSKS

Filter feeders and specialized carnivores on microorganisms, such as the scaphopods, are very difficult to maintain. Suspension feeding clams filter huge quantities of water daily. Unless you can continually add new seawater containing the microorganisms on which they feed, or unless they can adapt to easily cultured algae, brine shrimp, or other fine-grained fish food, you may have little success. Deposit feeders may survive quite well on organic debris and actually will help keep an aquarium clean. Clams need sufficient bottom sediments in which to burrow and lie in normal position, so that for infaunal clams, large containers are a necessity. Epifaunal (attached) clams can have the rock or shell to which they are attached placed on the aquarium bottom and may be kept more easily.

Intertidal dwelling snails present a special problem. They are accustomed to being exposed out of water by the falling tide, and then covered later by the rising tide. Dorothy Raeihle's article "Tabletop Marine Aquariums" in *How to Collect Shells* gives many tips on keeping marine snails. She points out the need to have porous rocks partly in and partly out of water to provide climbing places and damp surfaces above the water level. Without these surfaces, the snails may climb the wall of the aquarium, seal to the side, and dry out for lack of a "rising tide." Snails also may climb out of the water if the water has too little oxygen or is fouled by decaying matter. Daily checks that include pushing snails off the aquarium's side back into the water and continued aeration of the water will help greatly.

Food for herbivorous snails can be varied. Algae-covered rocks or corals can be kept in shallow trays exposed to sunlight, with a rock placed temporarily in the snail's aquarium for a day or so at regular intervals. A few marine species will eat lettuce or cabbage leaves, and some species will scrape a paste of cream of wheat or similar cereal that has been allowed to dry onto the glass wall just above the water level. Remember that many species feed only at times that are adjusted to the tide level or light levels. Many hobbyists have been frustrated by feeding the right food at the wrong time. Only experimentation will show what will and will not work with a particular species. Do not overfeed, and remove uneaten food after a few hours at most.

Carnivorous snails, unless specialized to feed on only one or two species, adapt well to bits of small clams, shrimp, or fat-free red meat. Carnivores can be fed less frequently than the herbivores, but it is even more essential that uneaten food be removed quickly and not allowed to decay and foul the water.

Regardless of your efforts, the diet provided probably will include only part of the vitamins, minerals, and other substances obtained by the mollusk in its natural habitat. Trace elements in the limited quantity of seawater will be extracted by the mollusk quickly, and then subsequent growth may be abnormal. Waste products released into the water by the mollusks may turn it slightly acid or otherwise react on the shell to cause erosion, pitting, or discoloration. Addition of clam shell fragments as an additional source of calcium may help alleviate these problems. "Instant seawater" packets may be adequate, but replacement by natural seawater from a nonpolluted area is preferred. New seawater should be filtered carefully through several layers of fine cloth before being added to the aquarium.

FRESHWATER MOLLUSKS

Species from a temporary spring pond or a stagnant woodland pool can stand lower levels of oxygen and heavier growths of algae in the water than species taken from a riffle area in a fast moving stream. The latter species need clear water and constant aeration by a mechanical pump. In all cases the freshwater aquarium should receive sunlight, or controlled ultraviolet light for plant growth, an hour or two each day. Intensive sunlight for a whole morning may cause too much plant growth to enable many snails and clams to survive in the aquarium.

Successful snail aquariums can be set up in 1 or 2 gallon tanks. These should be "planted" with bottom sediments and/or rocks, plants, or water from the same place that the mollusks were collected. Because large unionid clams can completely filter the water in a 2 gallon aquarium within a day or so, keeping clams alive can be very difficult. Some species will survive on fine fish food. Others can be kept only by using much larger tanks and exposing them to several hours of sunlight each day to encourage sufficient growth of algae and microscopic plants. As in all aquariums, the ideal is to establish a balance between plant growth, animal consumption, and bacterial reduction. When this is achieved, only water has to be added as a replacement for evaporation loss. Tap water should not be added directly, but allowed to sit in an open tray for a few days until chlorine and other added chemicals have diffused out.

Care must be taken not to crowd too many snails or fingernail clams into one tank, and the temperature of the water should be kept from fluctuating rapidly. In many parts of North America the water temperature of ponds, lakes, and rivers is normally far cooler than the temperatures inside a house. Species from these waters may require that the tank be chilled for best results.

Natural algae growths will normally provide enough food, provided the snails are not too crowded. Pellets of fish food or an occasional small piece of lettuce may be tried. The literature on home aquariums is comprehensive and well tested. Following their directions will usually be adequate for the care of a few freshwater snails.

LAND SNAILS AND SLUGS

Snails and slugs from moist forest litter are difficult to keep successfully. Snails from grassland areas or deserts are far easier to maintain in the home. The latter can be kept in glass-covered aquarium tanks with only a rock or two on which they can attach themselves. A piece of water-sprinkled lettuce or carrot slice will usually bring them to activity quite rapidly. Many arid zone species have a built-in biological rhythm. During a few months or weeks each year they will remain inactive and sealed up (aestivation) regardless of how much moisture and food is present in the terrarium. This period of inactivity corresponds with the dry season in their native land. Do not be alarmed if desert species remain sealed to a rock, ignoring both food and water for a month or two.

It is essential that the sides of the snail cage be washed periodically to get rid of slime trails and feces. As with marine and freshwater mollusks, removal of uneaten food is prudent.

Forest litter snails require more elaborate quarters. The floor of the cage should have a few screened over drain holes and the cage set on short feet in a tray to catch any seeping water. The holes will permit excess water to run out, since the snails need moisture, not soaking wet conditions. A layer of gravel or coarse stones should be covered by 2 or 3 inches of dirt from the forest. On top of the dirt, place leaves, wood fragments, and a piece or two of moss from the area in which the snails were collected. Above all, do not put too many snails in one cage. A rule of thumb would be to allow 35 cubic inches of terrarium for each snail that is 1 inch in diameter, 20 cubic inches for each snail that is $\frac{1}{2}$ inch in diameter, and proportionate amounts for smaller specimens. The top of the cage can be covered with fine mesh wire, whose holes should be smaller than the shell of newly hatched snails belonging to the species inside the terrarium. Otherwise there may be snail escapes. The surface litter should be sprinkled with water every other day. The soil must be moist, but not soaking weat. If the snails are kept in the upper floors of an average home or apartment, keeping adequate moisture levels will be a problem. But if the snails are kept in a damp basement, then shallow cake pans with only an inch of dirt inside can work as well as the more elaborate terrarium described above.

Frequent cleaning of the snail cages is mandatory. The sides should be wiped free of slime trails and feces once a week, the surface litter changed once every two weeks, and the soil replaced once a month. In changing the soil, be certain that you do not discard snail eggs, which many species bury in soil deliberately.

Despite these troubles, keeping land snails can be very enjoyable. A volunteer worker on my staff took home some live desert snails. She reported that the crunching noise made by these snails eating lettuce was pleasant company during midmorning coffee. Raw tomato, potato, dandelion greens, wet newspaper, cabbage, oatmeal, lettuce, and beans are eaten by many species. Carnivorous land snails will survive best if fed on snails that live in their native area. For short periods they can exist on the "edible snails" sold live in many ethnic markets of bigger cities.

Slugs are very messy in cages. Their copious slime production soon covers most surfaces and their surprising ability to slip through narrow openings may easily result in escapes from the cage to the great dismay of "malacophobic" relatives. Frequently slugs will feed on a much wider variety of foods than snails will, and slugs usually consume a greater quantity of food. Generally only two or three slugs should be kept in a 1 gallon aquarium tank. They should have a piece of moist wood or bark under which they can crawl and thus have a retreat from light and dryness. Damp paper toweling in another corner of the terrarium will provide a continuing source of moisture and emergency food.

Index

The references to illustrations are shown in bold face type. Scientific names, structures, scientists, general principles and selected topics have been listed, but geographic localities, references to most vertebrates, and functions or processes are omitted. Common names are cross-referenced to the scientific names.